STO

PRENTICE-HALL SERIES IN MATHEMATICAL ECONOMICS
Donald V. T. Bear, *Series Editor*

Dennis J. Aigner
  BASIC ECONOMETRICS

David A. Bowers and Robert N. Baird
  ELEMENTARY MATHEMATICAL MACROECONOMICS

Michael D. Intriligator
  MATHEMATICAL OPTIMIZATION AND ECONOMIC THEORY

Ronald C. Read
  A MATHEMATICAL BACKGROUND FOR ECONOMISTS

Menahem E. Yaari
  LINEAR ALGEBRA FOR SOCIAL SCIENCES

Linear Algebra
for
Social Sciences

MENAHEM E. YAARI

Associate Professor of Economics
The Hebrew University
Jerusalem, Israel

Linear Algebra
for
Social Sciences

PRENTICE-HALL, INC., ENGLEWOOD CLIFFS, N.J.

LINEAR ALGEBRA FOR SOCIAL SCIENCES
by Menahem E. Yaari

Printed in the United States of America

13-536839-1
Library of Congress Catalog Card Number 73-117207

Current printing (last digit)
10 9 8 7 6 5 4 3 2 1

PRENTICE-HALL INTERNATIONAL, INC., LONDON
PRENTICE-HALL OF AUSTRALIA PTY. LTD., SYDNEY
PRENTICE-HALL OF CANADA, LTD., TORONTO
PRENTICE-HALL OF INDIA PRIVATE LIMITED, NEW DELHI
PRENTICE-HALL OF JAPAN, INC., TOKYO

1566514

To My Mother

Series Foreword

The Prentice-Hall Series in Mathematical Economics is intended as a vehicle for making mathematical reasoning and quantitative methods available to the main corpus of the undergraduate and graduate economics curricula.

The Series has been undertaken in the belief that the teaching of economics will, in the future, increasingly reflect the discipline's growing reliance upon mathematical and statistical techniques during the past 20 to 35 years and that mathematical economics and econometrics ought not to be "special fields" for undergraduates and graduate students, but that every aspect of economics education can benefit from the application of these techniques.

Accordingly, the Series will contain texts that cover the traditional substantive areas of the curriculum—for example, macroeconomics, microeconomics, public finance, and international trade—thereby offering the instructor the opportunity to expose his students to contemporary methods of analysis as they apply to the subject matter of his course. The composition of the early volumes in the Series will be weighted in favor of texts that offer the student various degrees of mathematical background, with the volumes of more substantive emphasis following shortly thereafter.

As the Series grows, it will contribute to the comprehensibility and quality of economics education at both the undergraduate and graduate levels.

DONALD V. T. BEAR, *Series Editor*

Preface

In many universities today, departments of economics, industrial management, political science, psychology, and sociology offer their students a basic course in mathematics. Such a course is needed, to ensure that students will be adequately prepared for studying probability and statistical inference, linear and nonlinear programming, decision theory, learning theory, the theory of games, and other topics that involve mathematics and have become part and parcel of modern social science. One year of college calculus is no longer an adequate mathematical training for the social science student, so many social science departments have special courses that deal with such topics as set theory, linear algebra, convex sets, and advanced calculus. This book is intended to serve as a text in such a course. It covers only two of these topics, namely, elementary set theory (including a discussion of relations and functions) and linear algebra. However, it contains a chapter on linear inequalities, and linear inequalities usually come under the heading of convexity.

There are several texts in existence on mathematics for social scientists. Unfortunately, many of them tend to be based on the implicit notion that the social scientist's mathematics (in contrast with the mathematician's mathematics) must pay a price for its "practicality" in terms of reduced rigor and lack of precision. The main result of this tendency has been that students of the social sciences often study mathematics while gaining little or no familiarity with the basic modes of thinking that are involved in mathematical reasoning. If nothing else, this approach makes for a mathematical training that is piecemeal and rather uninviting. The present volume is based on the belief that a good way to make mathematical training attractive and long-lasting is to bring out clearly the fact that mathematics is primarily a study of the structure of certain abstract objects, and that behind the practical usefulness of mathematics in providing tools for scientific investigation lies a certain formal similarity between these abstract objects and the objects studied by the scientist. Thus, the main emphasis in this volume is on structures and on concepts. What makes this a book in mathematics *for social sciences*, then, is just the choice of topics to be covered. I have attempted to determine which topics to include, and which to exclude, by the criterion of usefulness to the social science student.

The present volume is based upon a set of lecture notes. I used these notes five times in the past few years, both with first-year graduate students and with advanced undergraduates. Judging by this experience, I would say that the difficulties that the students might expect to encounter in a course based on this book are somewhat akin to those encountered in taking up a foreign language. It turns out that the primary task of the instructor, especially during the first few weeks of the course, is to help his students overcome the bewilderment brought about by a multitude of new concepts. Once a student grasps the new concepts and gets a feel for them, he is able to follow the line of reasoning that is built on these concepts without too much difficulty.

At the end of each chapter there is a set of problems. Each set is designed to serve as a possible homework assignment. My experience has been that the average amount of time required to do a set of problems is four to five hours. In most cases, the problems, like the book itself, emphasize conceptual rather than technical matters. The idea is to force the student to think in terms of the "language" that he is studying. Some instructors will, no doubt, prefer to substitute other homework problems for those in the text.

It is indeed a pleasant duty to acknowledge my debt to friends and colleagues who have commented on early drafts of this volume. In particular, I would like to thank Professor Yitzhak Katznelson of the Hebrew University, Professor Donald Bear of the University of California at San Diego, and Professor Dennis Smallwood of Princeton University.

M.E. YAARI

Contents

Linear Algebra
for
Social Sciences

Part I BASIC CONCEPTS

1 Sets

In the present volume, the notion of a **set** and the notion of **belonging to a given set** will both be treated as undefined (or primitive) notions like, say, the notion of a point in Euclidean geometry. We therefore agree that a phrase such as "individual a belongs to the set A" is a meaningful sentence (i.e., it is either true or false), even though it contains words for which we do not have prior definitions.

The reader may be familiar with the fact that the notions of a set and of belonging to a given set, if left to their own devices, may lead to serious difficulties, having to do with sets which belong to themselves. The reference here is to *Russell's paradox*, which may be summarized as follows: Let a set be called *normal* if (and only if) it does not belong to itself. Thus, for example, the set of all chairs is normal, but the set of all sets is not. Now consider the set of all normal sets. Suppose first that this set is normal. Then, it must belong to the set of all normal sets, that is, it must belong to itself and therefore, by definition, it is *not* normal. On the other hand, suppose that the set is not normal. Then, it belongs to itself. This means that it belongs to the set of all normal sets and it must therefore be one of these normal sets. Thus, either hypothesis leads to a contradiction.

In order to avoid this predicament, let us agree in advance that the phrase "the set A belongs to the set A" shall remain *meaningless* for every set A.

Under this agreement, Russell's paradox does not arise. Note, however, that this is achieved at the cost of restricting the language somewhat. (For example, our language will not contain the phrase "the set of all sets.") In some contexts, this restriction of the language is too severe to be considered acceptable, so other ways of resolving the paradox must be sought. In the present context, however, the restriction will do no harm.

The symbol $\in$ will be used to denote "belongs to." Thus, $x \in X$ means "item x belongs to the set X." When $x \in X$, one says also that x is a **member** of X or that x is an **element** of X. Whenever the members of a set are spelled out, they are enclosed in braces. Thus, $\{1, 2, 3\}$ is the set whose only members are the integers 1, 2, and 3. Three dots $(\ldots)$ are often used to help "spell out" the elements of a set when there are infinitely many of them. For example, the formula $\{1, 2, 3, \ldots\}$ is generally accepted as proper notation for the set of all positive integers.

There is a close connection between sets and attributes (or predicates). With every attribute one may associate a set (the set of all individuals possessing the given attribute) and every set may be used to define an attribute (the attribute that is common to all members of the set exclusively.) It is therefore obvious that the notion of an attribute (or a predicate) is as much an undefined notion as that of a set. In any case, it is useful to have a nota-

3

tion for the set of all individuals possessing a given attribute. Let A be an attribute, and let Ax be the sentence "Individual x possesses the attribute A." We wish to define the set (call it S) of all individuals possessing the attribute A. The following notation is used to achieve this end:

$$S \underset{\text{def}}{=} \{x \mid Ax\}.$$

The symbol $\underset{\text{def}}{=}$ stands for "is by definition" and the vertical bar $\mid$ stands for the expression "such that." The formula above should therefore be read as follows: "S is, by definition, the set of all x such that x possesses attribute A."

Consider two sets, A and B. We say that the set A **contains** the set B (or that B **is contained in** A), and we write $A \supset B$, iff† every member of B is also a member of A. If $A \supset B$ then B is said to be a **subset** of A, which may also be written $B \subset A$. If $A \supset B$ and $A \subset B$ are both true, then the sets A and B are said to **coincide** or to be **equal** (some say **identical**) and one writes $A = B$. If $A \supset B$ is true, but $A \subset B$ is false, then B is said to be a **proper subset** of A. Clearly, every set is a subset of itself, but no set is a proper subset of itself.

The set of all elements that belong *either* to A *or* to B, or to both, is called the **union** of A and B and is denoted $A \cup B$. The set of all elements that belong *both* to A and to B is called the **intersection** of A and B, and is denoted $A \cap B$. The set of all elements that belong to A but do not belong to B is called the **complement of B with respect to A**, denoted $A \sim B$. Sometimes, when the set A is understood from the context, one may speak of the complement of B, denoted $\sim B$, without mentioning A explicitly.

It is convenient not to preclude the possibility of a set that has no members, i.e., of a set S for which the requirement

$$x \notin S \text{ for every } x \text{ satisfying } x \neq S$$

holds.‡ Such a set is called **empty**, or **null**. Furthermore, we now postulate that *if S is empty and A is any set, then $S \subset A$.*§ In other words, an empty set is a subset of every set. It now follows immediately that there is but one empty set. This set is called **the** empty set, and the symbol $\varnothing$ is used to denote it.

Two sets, A and B, are said to **overlap** iff they have a nonempty intersection (i.e., iff $A \cap B \neq \varnothing$). If they have an empty intersection ($A \cap B = \varnothing$), then they are said to be **disjoint**.

---

† Throughout this volume, the abbreviation *iff* will be used for *if and only if*.

‡ We write $x \notin S$ for the denial of the assertion $x \in S$.

§ This postulate actually follows from the definition of the symbol $\subset$, together with the properties of logical implication.

The set-theoretic symbols $\in$, $\subset$, $\supset$, etc., are used to state and prove various propositions, called theorems of set theory. To illustrate, let us state the following proposition, known as *De Morgan's law*: Consider two sets, A and B. Then,

(i) $\sim(A \cup B) = \sim A \cap \sim B$;

(ii) $\sim(A \cap B) = \sim A \cup \sim B$.

To prove these assertions, one must show that x belongs to the set that appears on the left-hand side iff it belongs to the set on the right-hand side. We leave the detailed proof to the reader. Here is another example of a simple proposition in set theory:

$$A \subset B \quad \textit{iff} \quad A \sim B = \varnothing.$$

Proof: The assertion $A \subset B$ is true iff, for every x, if $x \in A$ then $x \in B$. But this latter statement is true iff it is impossible to find an x that belongs to A but does not belong to B, i.e., iff the set $A \sim B$ is empty. Further examples of such propositions can be found in the problems at the end of the chapter.

Consider a set A, and suppose that the following two conditions hold: (1) A is nonempty. (2) If $x \in A$ and $y \in A$, then $x = y$. In this case, A is called a **one-element set** or a **singleton**. Now consider a set B, and suppose that the following three conditions hold: (1) B is nonempty. (2) B is not a singleton. (3) If $x \in B$, $y \in B$, and $z \in B$, then either $x = y$ or $x = z$ or $y = z$. Then, B is called a **two-element set**. Clearly, it is possible in this fashion to define an n-element set for arbitrary n, although the definitions quickly become very cumbersome.

One last remark about terminology: In the sequel, the words *family*, *class*, and *collection* will be used as synonyms for the word "set."

1.1. Prove that the empty set is unique (i.e., that there is only one empty set), and give a detailed proof of De Morgan's law.

1.2. In each of the following cases, write down the sets $A \cup B$, $A \cap B$, $A \sim B$, and $B \sim A$:

    (i) $A =$ any set
         $B = \varnothing$;

   (ii) $A = \{0, 1, 2\}$
        $B = \{1, 2\}$;

  (iii) $A =$ the set of positive integers
        $B =$ the set of negative integers.

1.3. Let A, B, and C be arbitrary sets. In each of the following cases, determine whether the given assertion is true or false. If it is true, prove it. If it is false, give a counterexample.

    (i) $A \cup (A \cap B) = A$;

   (ii) $A \sim (A \cap B) = (A \cup B) \sim B$;

  (iii) $A \sim (B \cup C) = (A \sim B) \cup (A \sim C)$;

(iv) $A \sim (B \cap C) = (A \sim B) \cap (A \sim C)$.

1.4. Write down the definition of a *three-element set*.

1.5. Prove the following assertion: A is a two-element set iff $A = B \cup C$, where B and C are disjoint singletons.

1.6. Let A and B be two sets. The **symmetric difference** of A and B, written $A \triangle B$, is defined by

$$A \triangle B \underset{\mathrm{def}}{=} (A \sim B) \cup (B \sim A).$$

Of the following assertions, which is true and which is false? (Supply proofs or counterexamples.)

(i) $A \triangle (A \cap B) = (A \cup B) \triangle B$;

(ii) $A \triangle (B \cup C) = (A \triangle B) \cup (A \triangle C)$;

(iii) $A \cap (B \triangle C) = (A \cap B) \triangle (A \cap C)$.

2 Relations

Consider two arbitrary nonempty sets, A and B. It is clearly possible to choose two "representatives," say, $a \in A$ and $b \in B$, and then form the set $\{a, b\}$. This is a two-element set about which we have the further information that a is taken from A and b is taken from B. The notation $\{a, b\}$ by itself does not convey this extra information. Often, what one does in this case is to agree that whenever reference is made to a set composed of an element from A and an element from B, the element from, say, A is written on the left and the element from B on the right. In addition, the braces, $\{\ \}$, must be abandoned in favor of some other kind of brackets, or else confusion is likely to arise. In the present volume, we shall use pointed brackets, $\langle\ \rangle$, for this purpose. We thus arrive at the notation $\langle a, b \rangle$ for the set $\{a, b\}$ with the further specification that $a \in A$ and $b \in B$.

Objects of the form $\langle a, b \rangle$, $\langle x, y \rangle$, etc. are called **ordered pairs**. One usually refers to a and b as, respectively, the **first component** and the **second component** of $\langle a, b \rangle$. Often, the identity of the sets from which these components are taken (such as the sets A and B of the previous paragraph) is not given explicitly but may be inferred from the context.

By virtue of the additional information embodied in them, ordered pairs differ from two-element sets in several respects. In the first place, it is clear that, while

$$\{a, b\} = \{b, a\}$$

we have, generally speaking,

$$\langle a, b \rangle \neq \langle b, a \rangle.$$

Note also that there is nothing wrong with an ordered pair of the form $\langle a, a \rangle$. For example, if $A = B = \{a\}$, then the *only* ordered pair that can be formed from "representatives" of A and B is the pair $\langle a, a \rangle$. In the case of sets, however, writing expressions of the form $\{a, a\}$ involves a redundancy, because $\{a, a\} = \{a\}$.

Given the notion of an ordered pair, we can proceed to use it to define an **ordered triple** as follows:

$$\langle a, b, c \rangle \underset{\text{def}}{=\joinrel=} \langle \langle a, b \rangle, c \rangle.$$

Similarly, we can now use the notion of an ordered triple and the notion of an ordered pair to define what we shall call an **ordered 4-tuple**:

$$\langle a, b, c, d \rangle \underset{\text{def}}{=\joinrel=} \langle \langle a, b, c \rangle, d \rangle.$$

9

In general, given that the ordered $(n-1)$-tuple $\langle a_1, a_2, \ldots, a_{n-1} \rangle$ is well-defined, we can define the **ordered n-tuple** $\langle a_1, a_2, \ldots, a_n \rangle$ by

$$\langle a_1, a_2, \ldots, a_n \rangle \underset{\text{def}}{=} \langle \langle a_1, a_2, \ldots, a_{n-1} \rangle, a_n \rangle.$$

a_1 will be referred to as the **first component** of $\langle a_1, a_2, \ldots, a_n \rangle$, a_2 will be referred to as the **second component** of $\langle a_1, a_2, \ldots, a_n \rangle$, and so on, up to a_n, which will be referred to as the **n-th component** of $\langle a_1, a_2, \ldots, a_n \rangle$.

It is, of course, possible to introduce the notion of an ordered n-tuple directly, without making use of the notion of an ordered pair. Let $\mathscr{F} = \{A_1, A_2, \ldots, A_n\}$ be a family of n nonempty sets. Choose an element from each of the sets $A_1, \ldots, A_n$ and then form the set of elements so chosen, taking care to write the representative of A_1 first, that of A_2 second, and so on. The result will be an ordered n-tuple. (We note in passing that if the family of nonempty sets $\mathscr{F}$ has infinitely many members, then the possibility of choosing a representative element from each set, and of forming the set of elements so chosen, must be postulated as an axiom, called the *axiom of choice*.)

We are now ready to define the notion of a relation. Consider two non-empty sets, A and B. A set of pairs† of the type $\langle a, b \rangle$, where $a \in A$ and $b \in B$, will be referred to as a **binary relation** between elements of A and elements of B. Similarly, consider n nonempty sets $A_1, A_2, \ldots, A_n$. A set of n-tuples of the type $\langle a_1, a_2, \ldots, a_n \rangle$, where $a_i \in A_i$ for $i = 1, 2, \ldots, n$, will be referred to as an **n-ary relation** among the elements of $A_1, A_2, \ldots, A_n$.

Thus we see that a relation is defined, essentially, by listing all its instances. A binary relation, for example, is defined by listing all the pairs such that the first component of the pair stands in this relation to the second component. Some people refer to the set of instances of a given relation as the **graph** of this relation. In this volume, however, a relation and its graph are one and the same thing.

Let R be a binary relation between the elements of a set A and the elements of a set B. The phrase "between the elements of A and the elements of B" is rather cumbersome, so in most cases we shall say, simply, that R is a relation between A and B. In cases where $A = B$, we shall say that R is a relation *in* the set A.

Here are some examples of relations:

1. Let $N = \{1, 2, 3, \ldots\}$ be the set of all positive integers. Define the set G by

$$G = \{\langle n, m \rangle \mid n \in N, m \in N, \text{ and } n = m + p \text{ for some } p \in N\}.$$

---

† From now on, we shall omit the adjective "ordered" from "ordered pair" and "ordered n-tuple."

Then, G is the binary relation ordinarily called "greater than . . ." in the set N.

    2. Let N be defined as in the previous example. Define R as follows:

$$R = \{\langle m, n, p\rangle \,|\, m \in N,\, n \in N,\, p \in N,\, m = 3p,\text{ and } n = 2p\}$$

In other words, R is the relation such that a triple $\langle m, n, p\rangle$ is an instance of R iff $m : n : p = 3 : 2 : 1$. Note that R is a trenary relation in the set N.

    3. Let $A = \{\text{Socrates, Plato, Aristotle}\}$ and let $B = \{\text{cheese, } 17.6\}$. Define the set Q by

$$Q = \{\langle\text{Socrates, } 17.6\rangle, \langle\text{Aristotle, cheese}\rangle\}.$$

Then, Q is a binary relation between A and B, even though in everyday life we are not very likely to find a sense in which we might say that Socrates is related to 17.6 in the same way that Aristotle is related to cheese.

    The sentence "a stands in relation R to b" is written in symbols as follows: $\langle a, b\rangle \in R$. However, whenever there is no danger of confusion, one often replaces $\langle a, b\rangle \in R$ with the more compact notation aRb. There is no similar abbreviation for a general n-ary relation.

    In view of the fact that ordered n-tuples can be looked upon as ordered pairs composed together, it is possible to regard n-ary relations as composite binary relations. For example, a trenary relation may be looked upon as a binary relation between two sets, one of which is itself a binary relation. This is one of the reasons for the fact that many discussions of relations concentrate mainly on binary relations. (Another reason is that higher order relations are just not very common.) Here too, we shall confine the remainder of the chapter to binary relations.

    Consider a binary relation R between two sets, A and B. The **domain** of R, denoted $\mathscr{D}(R)$, is the set of all members of A that stand in relation R to some member of B. The **counterdomain** of R, denoted $\mathscr{C}(R)$, is the set of all members of B for which there is a member in A standing in relation R to them. Symbolically,

$$\mathscr{D}(R) = \{x \,|\, x \in A \text{ and } xRy \text{ for some } y \in B\}$$
$$\mathscr{C}(R) = \{y \,|\, y \in B \text{ and } xRy \text{ for some } x \in A\}.$$

Clearly, $\mathscr{D}(R) \subset A$ and $\mathscr{C}(R) \subset B$.

    With every relation R between the sets A and B one may associate a new relation, to be denoted R^{inv}, between B and A, by postulating that $xR^{inv}y$ iff yRx. Thus,

$$R^{inv} = \{\langle x, y\rangle \,|\, yRx\}.$$

The relation R^{inv} is called the **inverse** of R (some say the **converse** of R). It is immediately clear that taking the inverse of R^{inv} brings us right back to R. In other words, the assertion

$$(R^{inv})^{inv} = R$$

is true for every binary relation R.

Consider once again two arbitrary sets, A and B. The set of all possible pairs $\langle a, b \rangle$, with $a \in A$ and $b \in B$, is called the **Cartesian product** of A and B, and it is denoted $A \times B$:

$$A \times B = \{\langle x, y \rangle \mid x \in A \text{ and } y \in B\}.$$

Similarly, given n arbitrary sets, $A_1, A_2, \ldots, A_n$, one defines the Cartesian product $A_1 \times A_2 \times \cdots \times A_n$ by

$$A_1 \times A_2 \times \cdots \times A_n = \{\langle x_1, \ldots, x_n \rangle \mid x_i \in A_i, i = 1, \ldots, n\}.$$

The Cartesian product $A \times B$ is clearly a relation between the sets A and B. It is, in fact, the most general relation between A and B, since every member of A stands in this relation to every member of B. For this reason, $A \times B$ is also referred to as the **universal** relation between A and B. Every relation between the sets A and B is, of course, a subset of $A \times B$. Indeed, the following is an alternative definition of a relation: A binary relation between the sets A and B is a subset of the Cartesian product $A \times B$. (A similar definition may be given for an n-ary relation.) Let us recall that the empty set $\varnothing$ is a subset of every set and, in particular, it is a subset of $A \times B$. Therefore, $\varnothing$ is (trivially) a relation between A and B. We may call it the *void* relation between A and B, since no member of A stands in this relation to any member of B.

An important class of relations is that in a given set A—i.e., relations that hold between members of the set A and members of the same set. Such relations are often classified according to whether or not they possess certain properties. The following four properties are the most important:

1. *Reflexivity:* A relation R in a set A is said to be **reflexive** iff xRx for all $x \in A$; R is said to be **irreflexive** iff xRx is false for every $x \in A$. If R is neither reflexive nor irreflexive, then it is said to be **nonreflexive**.

2. *Symmetry:* A relation R in a set A is said to be **symmetric** iff for every $x \in A$ and $y \in A$, xRy implies yRx. R is said to be **asymmetric** iff for every $x \in A$ and $y \in A$, xRy excludes yRx. Furthermore, R is said to be **antisymmetric** iff for every $x \in A$ and $y \in A$, if xRy and yRx are both true, then $x = y$. (Equivalently, R is antisymmetric iff it is never

true that xRy and yRx and $x \neq y$.) A **nonsymmetric** relation is one that is neither symmetric nor asymmetric, nor antisymmetric.

Three things may be noted: First, if R is asymmetric then R is antisymmetric. Second, if R is asymmetric then R is irreflexive. Finally, if R is both antisymmetric and irreflexive, then R is asymmetric. Thus, asymmetry is equivalent to the conjunction of antisymmetry and irreflexivity.

3. *Transitivity:* A relation R in a set A is said to be **transitive** iff for every $x \in A$, $y \in A$ and $z \in A$, xRy and yRz together imply xRz. A relation that is not transitive is referred to as **nontransitive**. Some authors define R to be an **intransitive** relation iff xRy and yRz together exclude xRz.

4. *Connectedness:* A relation R in a set A is said to be **connected** iff for every $x \in A$ and $y \in A$, if $x \neq y$ then either xRy or yRx (or both).

A rather simple relation in a set A is the so-called **identity relation**, which will be denoted I_A, and which is defined by

$$I_A = \{\langle x, x \rangle \,|\, x \in A\}.$$

Note that I_A is reflexive, symmetric, transitive, but not connected if A has more than one member. I_A is not the only relation having this particular combination of properties.

EQUIVALENCE RELATIONS

Let R be a relation in a set A. If R is reflexive, symmetric, and transitive, it is referred to as an **equivalence relation** in A. For example, the relation "born in the same year as . . ." is an equivalence relation in the set of human beings. Let a be an element of A. If R is an equivalence relation in A, then the set $\{x \,|\, x \in A \ and \ xRa\}$ is called the **R-equivalence class** of a in A. For example, if R is the relation "born in the same year as . . ." and A is the set of all human beings, then the R-equivalence class of George Washington is simply the set of all human beings born in the year 1732. If an equivalence relation is defined in A, then it is possible to partition A into a family of disjoint equivalence classes.

ORDERINGS AND QUASI ORDERINGS

Consider once again a relation R in a set A. If R is both transitive and antisymmetric, then it is called an **ordering** of the set A. The ordering R is said to be **complete** if, in addition to being transitive and antisymmetric, R is also connected in A. An ordering that is not complete is sometimes called a **partial** ordering. To illustrate, let S be an arbitrary set, and let $\mathscr{F}$ be the family of all subsets of S. Then, the relation $\subset$ (i.e., the relation "is a subset of . . .") is a partial ordering of the family $\mathscr{F}$, for $\subset$ is transitive and antisymmetric, but it is not connected (if S has more than one member). Family trees also provide good examples of partial orderings. A family tree may be thought of as a graphic representation of the relation "is a descendant of . . ." and this relation is transitive, antisymmetric, but not connected (unless the family tree has but one branch). For an example of a complete ordering, we turn to the set N of all the positive integers. The relation "greater than . . ." (which is denoted $>$) is a complete ordering of N, as may readily be verified.

Let R be an ordering. If, in addition to its other properties, R is irreflexive, then it is referred to as a **strict** ordering. (Recall that a relation is both irreflexive and antisymmetric iff it is asymmetric.) Returning for a moment to the examples of the previous paragraph, we find that the last two ("is a descendant of . . ." and "is greater than . . .") are examples of a strict ordering, while the first ("is a subset of . . .") is not. In general, nonstrict orderings are easily convertible to strict orderings, and vice versa. For example, the relation "is a proper subset of . . ." is a strict ordering, while the relation "is greater than or equal to . . ." is a nonstrict ordering.

A relation that is transitive but not antisymmetric may nevertheless be used on occasion to order a set. As an ordering, however, such a relation is in general weaker than the orderings discussed above. Formally, if a relation R in a set A is transitive and reflexive, then R is said to be a **quasi ordering**, or a **preordering**, of A. (Some authors refer to such a relation as a **weak ordering**.) Note that every equivalence relation is a quasi ordering, which serves to illustrate how weak are the order properties of a quasi ordering. If R is transitive, reflexive, and connected, it is referred to as a **complete quasi ordering**. Complete quasi orderings are prominent in economics. The theory of consumer's choice usually starts out with the definition of a so-called "preference-or-indifference" relation, which is a complete quasi ordering of the sets of states in which a consumer unit can find itself. If R is a complete quasi ordering of a set A, and if a is an element of A, then we define the **R-equivalence class** of a as the set

$$\{x \mid xRa \text{ and } aRx\}.$$

This set possesses all the elements of A that are in a *tie* with a in the ordering R. It is easy to verify that if R is a complete quasi ordering, and if Q is the relation defined by xQy iff xRy and yRx, then Q is an equivalence relation. The equivalence classes defined above are really Q-equivalence classes, but it has become common to refer to them also as R-equivalence classes.

PROBLEMS

2.1. Let R be a relation in a set A. Determine, among the following assertions, which is true and which is false:

(i) If R is reflexive, then $\mathscr{D}(R) = \mathscr{C}(R)$;

(ii) If R is symmetric, then $\mathscr{D}(R) = \mathscr{C}(R)$;

(iii) If R is transitive, then $\mathscr{D}(R) = \mathscr{C}(R)$;

(iv) If R is connected, then $\mathscr{D}(R) = \mathscr{C}(R)$.

In cases where you find the given assertion to be true, is the further assertion $\mathscr{D}(R) = \mathscr{C}(R) = A$ also true? Give proofs or counterexamples.

2.2. Let R be a relation in a set A. Define a new relation, R^2, also in A, by the requirement: For every $x \in A$ and $y \in A$,

$$x R^2 y \text{ iff } xRz \text{ and } zRy \text{ for some } z \in A.$$

Give necessary and sufficient conditions for R to be reflexive, for R to be symmetric, and for R to be transitive, in terms of the relations R^2, R^{inv} (the inverse of R) and I_A (the identity), and in terms of the set-theoretic symbols $\subset$, $\supset$, and $=$.

2.3. Let R be a complete quasi ordering. Define a relation $\tilde{R}$ by the requirement

$$x \tilde{R} y \text{ iff } not \ xRy.$$

What kind of a relation is $\tilde{R}$? Prove your assertion.

2.4. Let R be a complete quasi ordering and let $\mathscr{F}$ be the family of all R-equivalence classes. Define a relation R' in $\mathscr{F}$ by the following requirement: For every $A \in \mathscr{F}$ and $B \in \mathscr{F}$,

$$A R' B \text{ iff } aRb \text{ for some } a \in A, b \in B.$$

What kind of a relation is R'? Prove your assertion.

2.5. Let N be the set of all positive integers, and let m and n be members of N. We say that m is a **divisor** of n iff $n = pm$ for some $p \in N$. Define a relation R in N by

$$mRn \text{ iff } m \text{ is a divisor of } n.$$

What kind of a relation is R? Prove your assertion.

2.6. Let R be an equivalence relation in a set A, and let Q be an equivalence relation in a set B. Is $R \cup Q$ an equivalence relation in $A \cup B$? Prove your assertion.

2.7. Let R and Q be two complete strict orderings of a set A. Show that, if $Q \supset R$, then $Q = R$.

3 Functions

A binary relation f between two sets, A and B, is said to be a **function on A to B** (some say **from** A to B) iff the following two conditions hold:

(i) $\mathscr{D}(f) = A$;

(ii) For every $x \in A$, $y \in B$, and $z \in B$, if $\langle x, y \rangle \in f$
      and $\langle x, z \rangle \in f$, then $y = z$.

Let a be an arbitrary element of A. Then, condition (i) guarantees that there is *at least* one element in B standing in relation f to a, and condition (ii) guarantees that there is *at most* one such element in B.

Let f be a function on A to B. The domain of f, which is required by condition (i) to coincide with the set A, is also called the **domain of definition** of f. The counterdomain of f is invariably referred to as the **range of values** of f (or simply as the **range** of f, for short) and it is denoted $\mathscr{R}(f)$. If, perchance, $\mathscr{R}(f) = B$, then we say that f is a function on A **onto** B (or simply that f is **onto**).

The words *transformation* and *mapping* are used by most people as synonyms for the word "function." A function f on A to B is sometimes said to **map** A into B (or to map A onto B, if f is onto).

If f is a function on A to B, and if $\langle a, b \rangle \in f$, then one writes

$$b = f(a)$$

and one refers to b as the **value** of f at a. It is also common to say that f **carries** a **into** b, or that b is the **image** of a **under** f.

Consider a function f on A to B, and let C be a subset of A. The set

$$\{y \mid y = f(x) \, for \, some \, x \in C\}$$

is called the **image** of C **under** f, and it is denoted $f(C)$. Similarly, if D is a subset of B, then the set

$$\{x \mid y = f(x) \, for \, some \, y \in D\}$$

is called the **inverse image** of D **under** f and it is denoted $f^{-1}(D)$. A question now arises: Is there a function g on B to A such that $g(D) = f^{-1}(D)$ for every subset D of B? It is easy to verify that the answer to this question is in the affirmative iff the inverse relation of f, f^{inv}, is a function on B to A, in which case f^{inv} is the desired function g. The reader may wish to write down a detailed proof for this assertion (problem 3.1 at the end of the chapter), as this provides a useful exercise in the manipulation of sets. If the inverse relation of a function f on A to B is, in fact, a function on B to A, then this inverse is denoted f^{-1}, and it is referred to as the **inverse function** of f. If the inverse function f^{-1} exists, we say that the function f is **invertible**.

Note that there cannot be any question of ambiguity in the phrase "*the inverse function of f*" because the inverse relation associated with a given relation is unique, and therefore so is the inverse function associated with a given function.

In order to state the following proposition, we need the concept of a one-to-one function: A function f on A to B is said to be **one-to-one** iff for every $x \in A$ and $y \in A$,

$$f(x) = f(y) \text{ implies } x = y.$$

Let f be a function on A to B. Then, f is invertible iff f is one-to-one and onto.

PROOF:

Take the inverse relation of f, f^{inv}. The function f is invertible iff f^{inv} satisfies the two conditions in the definition of a function. Now, f^{inv} satisfies condition (i) iff f is onto, and f^{inv} satisfies condition (ii) iff f is one-to-one.

It is important to avoid confusing inverse images with inverse functions. (Unfortunately, the conventional notation is not very helpful in this respect.) Let us therefore reiterate: If f is a function on A to B, then the inverse image, $f^{-1}(D)$, is well defined for every subset D of B, regardless of whether f is invertible or not. On the other hand, the inverse function f^{-1} is well-defined iff f is invertible. (If f^{-1} exists, then $f^{-1}(D)$ is *both* the inverse image of D under f and the image of D under f^{-1}.)

If f is an invertible function, then it is immediately clear that f^{-1} is also invertible and that $(f^{-1})^{-1} = f$.

Note that I_A, the identity relation in a set A, which is defined by

$$I_A = \{\langle x, x\rangle \mid x \in A\},$$

is clearly a function on A to A. It is, furthermore, an invertible function, and its inverse is equal to itself:

$$(I_A)^{-1} = I_A \text{ for every set } A.$$

From now on, we shall refer to I_A as the **identity function** on A.

We turn now to the concept of the composition of two functions. Let

f be a function on A to B, and let g be a function on B to C. The **composition** of f and g, to be denoted $g \circ f$, is defined by the requirement

$$(g \circ f)(x) = g(f(x)) \text{ for every } x \in A.$$

Clearly, $g \circ f$ is a function on A to C. The direct definition of $g \circ f$, in terms of a set, is

$$g \circ f = \{\langle x, z\rangle \mid x \in A, z \in C, \text{ and } \langle x, y\rangle \in f, \langle y, z\rangle \in g \text{ for some } y \in B\}.$$

The following are four assertions involving the composition of functions:

1. If f, g, and h are functions for which the compositions $g \circ f$ and $h \circ g$ exist, then

$$h \circ (g \circ f) = (h \circ g) \circ f.$$

2. If f is a function on A to B, then

$$f \circ I_A = f \quad \text{and} \quad I_B \circ f = f.$$

3. If f is an invertible function on A onto B, then $f^{-1}(f(x)) = x$ for all $x \in A$, and $f(f^{-1}(x)) = x$ for all $x \in B$. In other words,

$$f^{-1} \circ f = I_A \quad \text{and} \quad f \circ f^{-1} = I_B.$$

4. If f, g, and h are functions for which the compositions $g \circ f$ and $h \circ f$ exist, and if f is invertible, then

$$g \circ f = h \circ f \quad \text{implies} \quad g = h.$$

Assertions 1, 2, and 3 follow at once from the relevant definitions, and assertion 4 follows from 1, 2, and 3.

Let f be a function on A to B, and let g be a function on B to C. If both f and g are invertible, then the composition $g \circ f$ is also invertible. Furthermore, we have

$$(g \circ f)^{-1} = f^{-1} \circ g^{-1}.$$

To see this, simply write out the set that defines the composition $f^{-1} \circ g^{-1}$, This set turns out to be equal to the *inverse relation* of $g \circ f$. Now, $f^{-1} \circ g^{-1}$ is a perfectly well-defined function on C to A. Thus, the inverse relation of $g \circ f$ is a function (i.e., $g \circ f$ is invertible) and, furthermore, this function is equal to $f^{-1} \circ g^{-1}$.

So far, we have dealt with the broad notion of a function on an arbitrary set A to an arbitrary set B. Nothing was assumed known about A and B, except that they are sets. In many instances, however, the sets A and B have very definite structures. If the structure of the set A and the structure of the set B are similar in nature, then it becomes of great mathematical interest to investigate functions on A to B that have the property of being *structure-preserving*. The meaning of the word "structure" here is admittedly rather vague, and deliberately so. The reason for this vagueness is that in different discussions, very different kinds of structure arise. For example, in linear algebra one is interested (as we shall see later in this volume) in investigating functions that preserve a certain algebraic structure. We shall refer to such functions in Chapter 5 as linear transformations. In analysis, on the other hand, one is often interested in functions that preserve a certain metric structure (or, more specifically, in functions that preserve *proximity*), and these are known as continuous functions. In general, if a function f on A to B is in some sense structure-preserving, then it is common to refer to f as a **homomorphism**. Clearly, the precise meaning of the word "homomorphism" depends on the meaning of the word "structure," and it will therefore vary from context to context.

The notion of an order-preserving function provides a good example of a homomorphism: Let f be a function on A to B. Assume also that R is an ordering of the set A and that Q is an ordering of the set B. The function f is called **order-preserving** with respect to R and Q iff for every $x \in A$ and $y \in A$,

$$xRy \quad \text{implies} \quad f(x)Qf(y).$$

For example, let N be the set of all positive integers, and let f be the function on N to N such that $f(x) = x^2$ for every $x \in N$. Then, f is order-preserving with respect to the relation "greater than ..." since $x > y$ implies $x^2 > y^2$. The term "order-preserving" is used also when the relevant relations are quasi orderings rather than orderings.

Before closing this chapter, we turn briefly to the description of a specific and rather common type of function. A function on the Cartesian product $A \times A$ to A is called a **binary operation** in A. Let $\triangle$ be a binary operation in A, and let x, y, and z be members of A such that $\langle x, y, z \rangle \in \triangle$. Then, instead of writing $z = \triangle(x, y)$, or $z = \triangle(\langle x, y \rangle)$, one sometimes writes

$$z = x \triangle y.$$

This notation is most prevalent in those cases where a special symbol, such as $+$, $\div$, $\cdot$, etc., is used to denote the binary operation.

Binary operations are often assumed to possess certain properties, of which the following are the most important:

1. A binary operation $\triangle$ in a set A is said to be **associative** iff for every $x \in A$, $y \in A$, and $z \in A$,

$$(x \triangle y) \triangle z = x \triangle (y \triangle z).$$

2. A binary operation $\triangle$ in A is said to be **commutative** iff for every $x \in A$ and $y \in A$,

$$x \triangle y = y \triangle x.$$

3. A binary operation $\triangle$ in A is said to satisfy the **law of cancellation on the left** iff for every $x \in A$, $y \in A$, and $z \in A$,

$$x \triangle y = x \triangle z \quad \text{implies} \quad y = z.$$

Similarly, $\triangle$ is said to satisfy the **law of cancellation on the right** iff

$$y \triangle x = z \triangle x \quad \text{implies} \quad y = z.$$

4. Finally, let $\triangle$ and $\otimes$ be two binary operations in a set A. We say that $\otimes$ is **left-distributive over** $\triangle$ iff for every $x \in A$, $y \in A$, and $z \in A$,

$$x \otimes (y \triangle z) = (x \otimes y) \triangle (x \otimes z).$$

As might be expected, $\otimes$ is called **right-distributive** over $\triangle$ iff for every $x \in A$, $y \in A$, and $z \in A$,

$$(x \triangle y) \otimes z = (x \otimes z) \triangle (y \otimes z).$$

If $\otimes$ is both left-distributive and right-distributive over $\triangle$, then we say simply that $\otimes$ is **distributive** over $\triangle$.

Examples of binary operations satisfying these properties will appear in the following chapters.

PROBLEMS

3.1. Consider a function f on A to B, and prove the assertion that a function g on B to A, satisfying the requirement

$$g(D) = f^{-1}(D) \text{ for every subset } D \text{ of } B,$$

exists iff the inverse relation f^{inv} is a function on B to A. (Remember that the implication must be proved both ways.)

3.2. Once again, let f be a function on A to B, and let g be a function on B to A. Are the following assertions true?

  (i) If $g \circ f = I_A$, then f is invertible and $f^{-1} = g$;
  (ii) If $f \circ g = I_B$, then f is invertible and $f^{-1} = g$.

Give proofs or counterexamples.

3.3. Under what conditions is the Cartesian product $A \times B$ a function on A to B?

3.4. Consider two sets, A and B, and assume that R is an ordering of A and that Q is an ordering of B. Let f be a function on A onto B that is order-preserving with respect to R and Q. Find conditions on R and Q which imply that f is an invertible function. Prove your assertion.

22

3.5. Let A be a finite set (that is, a set possessing a finite number of elements). A binary operation $\triangle$ in A is given. Under what conditions is the function $\triangle$ invertible? Prove your assertion. Would you change your answer if A were an infinite set?

3.6. Let Z be the set of all integers (the positive integers, the negative integers, and 0). Let a binary operation $*$ be defined in Z, and assume that the following two conditions hold:

    (i) $x * 0 = x$ for every $x \in Z$;

    (ii) The binary operation $+$ (ordinary addition) in Z is distributive over $*$.

Are these facts sufficient to determine the operation $*$? Prove your assertion.

4 The Real Numbers

Essential to much of the discussion in subsequent chapters will be the *real number system*, which will be denoted **R**. A step-by-step development of the real numbers, starting from the positive integers, and proceeding through the integers and the rational numbers, would carry us far beyond the scope of the present volume. Thus, we must be content simply to write down the properties that characterize the real numbers and to discuss them briefly.

Consider a nonempty set F, and assume that two binary operations, $+$ and $\cdot$, are defined on $F \times F$ to F. The operation $+$ is called **addition** and the operation $\cdot$ is called **multiplication**. The set F, together with the operations $+$ and $\cdot$, is assumed to satisfy certain axioms. We start with axioms about addition:

AXIOM 1

ADDITION IS ASSOCIATIVE.

AXIOM 2

ADDITION IS COMMUTATIVE.

AXIOM 3

THERE EXISTS IN F AN ELEMENT, DENOTED 0, WITH THE PROPERTY THAT $x + 0 = x$ FOR EVERY $x \in F$.

Before proceeding to the next axiom, it is necessary to show that the element 0, whose existence is postulated in Axiom 3, is unique. Without this fact, Axiom 4 below does not make any sense. To see that 0 is indeed unique, suppose that there are two elements in F, say a and a', such that (i) $x + a = x$ for every $x \in F$, and (ii) $x + a' = x$ for every $x \in F$. Then, in particular, we may set $x = a'$ in (i) and $x = a$ in (ii). This leads to $a' + a = a'$ and $a + a' = a$, respectively. But, by Axiom 2, $a' + a = a + a'$. We therefore conclude that $a = a'$.

AXIOM 4

ASSOCIATED WITH EACH ELEMENT x OF F IS AN ELEMENT OF F, DENOTED $-x$, SUCH THAT $x + (-x) = 0$.

It is easy to show that $-x$, the element of F associated with a given $x \in F$ in the manner described in this axiom, is unique. For suppose that there exist three elements of F, say, a, b, and c, such that $a + b = 0$ and $a + c = 0$. By Axiom 1, $(c + a) + b = c + (a + b)$. Now, $a + b = 0$ by hypothesis, and $c + a = 0$ by hypothesis and by Axiom 2. So, we have $0 + b = c + 0$, and it follows from Axioms 3 and 2 that $b = c$.

Corresponding to the four axioms on addition, we now have four axioms on multiplication:

AXIOM 5

MULTIPLICATION IS ASSOCIATIVE.

AXIOM 6

MULTIPLICATION IS COMMUTATIVE.

AXIOM 7

THERE EXISTS IN F AN ELEMENT, DENOTED 1, WITH THE PROPERTY THAT $x \cdot 1 = x$ FOR EVERY $x \in F$.

Once again, there can be only one element in F with the property described in Axiom 7. In other words, the element 1 is unique. It is therefore legitimate to write:

AXIOM 8

ASSOCIATED WITH EACH ELEMENT x OF F, *except* POSSIBLY FOR THE ELEMENT 0, IS AN ELEMENT OF F, DENOTED x^{-1}, SUCH THAT $x \cdot x^{-1} = 1$.

Here again, it is simple to show that x^{-1}, whenever it exists, is unique. The reason for excluding the element 0 in this axiom is investigated in problem 4.3, at the end of the chapter.

Finally, we have an axiom governing the *interaction* of addition and multiplication:

AXIOM 9

MULTIPLICATION IS DISTRIBUTIVE OVER ADDITION.

Every nonempty set F, in which two binary operations are defined so that Axioms 1–9 are satisfied, is called a **field**. There are many examples of fields, the real numbers being one of them. Every singleton is (trivially) a field, when addition and multiplication are defined in the only way possible.† The rational numbers and the complex numbers are fields, but the integers are not. (Why?) Some of the properties of fields will be explored in the problems at the end of the chapter.

The elements of a field are often referred to as **scalars**. Given a field F,

---

† Some authors make it one of the requirements defining a field that it possess at least two distinct members. Here, we shall not impose this requirement.

we can define **subtraction** and **division** in an obvious manner: If x and y are scalars, then

$$x - y \underset{\text{def}}{=} x + (-y)$$

$$x/y \underset{\text{def}}{=} x \cdot y^{-1}, \text{ provided } y^{-1} \text{ exists.}$$

In most of the subsequent chapters, only the field properties of the real numbers will be required. However, since further properties of the real numbers will, nevertheless, be needed occasionally, let us take a brief look at the remaining axioms for the real numbers. In all, there are five more axioms, of which the first four have to do with the introduction of an *ordering* into the field F, and with the interaction of the binary operations $+$ and $\cdot$ with this ordering.

AXIOM 10

THERE EXISTS A NON-VOID COMPLETE STRICT ORDERING $>$ (READ: GREATER THAN . . .) OF F.

The fact that $>$ is non-void and strict implies that F cannot be a singleton. An asymmetric non-void relation cannot be defined in a set F if F is a singleton.

AXIOM 11

IF x, y, AND z ARE MEMBERS OF F, THEN $x > y$ IMPLIES $x + z > y + z$.

AXIOM 12

FOR EVERY $x \in F$, $y \in F$, AND $z \in F$, $x > y$ AND $z > 0$ TOGETHER IMPLY $x \cdot z > y \cdot z$.

AXIOM 13

FOR EVERY $x \in F$, $y \in F$, AND $z \in F$, $x > y$ AND $0 > z$ TOGETHER IMPLY $y \cdot z > x \cdot z$.

Using the ordering $>$, one can define three related orderings of F, $\geq$, $<$, and $\leq$, as follows:

$$x \geq y \quad \text{iff} \quad x > y \quad \text{or} \quad x = y,$$
$$x < y \quad \text{iff} \quad y > x,$$
$$x \leq y \quad \text{iff} \quad y > x \quad \text{or} \quad y = x.$$

The set F, with the operations $+$ and $\cdot$, and with the relation $>$, is said to be an **ordered field**, whenever Axioms 1–13 hold.

In addition to the real numbers, the rational numbers are also an ordered field, but the complex numbers are not. In order to characterize the real numbers completely, we need one more axiom.

Let F be an ordered field, and let G be a nonempty subset of F. An element a of F is said to be an **upper bound** of G iff $a \geqq x$ for every $x \in G$. If the set G has an upper bound, we say that G is **bounded above**. Finally, an element a of F is said to be a **least upper bound** of G iff a is an upper bound of G and $a' \geqq a$ whenever a' is also an upper bound of G.

AXIOM 14

LET G BE A NONEMPTY SUBSET OF F. IF G IS BOUNDED ABOVE, THEN G HAS A LEAST UPPER BOUND.

It is Axiom 14 that separates the real numbers from other ordered fields. Indeed, we may now define the **system of real numbers** as a set F, ordered by a relation $>$, and in which there are defined two binary operations, $+$ and $\cdot$, in such a way that Axioms 1–14 are satisfied.

In order to be able to speak about **the** system of real numbers, one must show that every system that satisfies Axioms 1–14 can be converted into any other system satisfying the same axioms simply by changing the names of the elements, so that, except for a possible difference in names, the real number system is unique. Without delving deeply into this matter, let us just note that as long as one takes care to use only those properties of the real numbers that are postulated in Axioms 1–14, there is no way in which differences between two systems, both of which satisfy Axioms 1–14, can come into play. This is, roughly speaking, the argument that permits us to speak about the real number system being uniquely determined by Axioms 1–14.

As has already been mentioned, we shall use the symbol **R** to denote the real number system. The assertion "a is a real number" will be written $a \in$ **R**, even though **R** is more than just a set.

Within the real number system **R**, one can identify subsets corresponding to the positive integers, to the integers, and to the rational numbers. For example, let us define a subset N of **R** by the following two requirements:

(i) $1 \in N$;
(ii) if $x \in N$, then $(x + 1) \in N$.

Then, N is the set of positive integers, considered a subset of the real numbers. Of course, in order to *prove* that N is the set of positive integers, we must first write down the axioms for the positive integers, and then show that our set, N, satisfies these axioms. There are similar discussions in the cases of the integers and the rational numbers. In the present volume, these discussions will be omitted, because they would carry us too far afield.

We close the chapter with a remark on terminology: A function whose range of values is contained in **R** is called a **real valued function**. Some authors use the term **real function** for a function whose domain and range are both contained in **R**.

PROBLEMS

4.1. Show that, in a field, addition satisfies the law of cancellation (on the left as well as on the right).

4.2. Let F be a field. Prove the following assertions:
   (i) $x \cdot 0 = 0$ for every $x \in F$;
   (ii) $x \cdot (-1) = -x$ for every $x \in F$.

4.3. Consider the axioms that define a field. Show that if 0 is not excluded in Axiom 8, then the "field" defined by Axioms 1–9 is a singleton.

4.4. If a field has more than one element, then it has infinitely many elements. Is this assertion true? Prove your answer.

4.5. Let F be an *ordered* field, and consider the following four assertions:
   a. There exists a subset P of F, such that for every $x \in F$, if $x \neq 0$ then either $x \in P$ or $-x \notin P$.
   b. For every $x \in F$, if $x \in P$ then $-x \notin P$.
   c. For every $x \in F$, $y \in F$, if $x \in P$ and $y \in P$, then $(x + y) \in P$.
   d. For every $x \in F$, $y \in F$, if $x \in P$ and $y \in P$, then $(x \cdot y) \in P$.
Show that assertions A–D are, together, *equivalent* to Axioms 10–13.

4.6. Show that in the real number system, $x = -x$ implies $x = 0$. Is this implication valid in every field?

4.7. Let a be a nonzero element of a field. Show that

$$(-a)^{-1} = -a^{-1}.$$

Part II LINEAR
ALGEBRA

5 Real

Linear Spaces

In order to construct a linear space, we need a nonempty set, say L, and a field, say F. Recall that F has two binary operations, addition and multiplication, defined in it, and that these binary operations satisfy certain conditions. These conditions were listed as Axioms 1–9 in the previous chapter. We begin by defining a binary operation in L, which will also be called **addition** and denoted $+$. We shall have to take care not to confuse addition in L with addition in F. In order to help keep the two apart, let us use lowercase Greek letters (α, β, etc.) for members of F and lowercase Roman letters (in italics) for members of L.† Addition in L is assumed to satisfy the same four axioms that are satisfied by addition in F:

AXIOM 1

ADDITION IN L IS ASSOCIATIVE.

AXIOM 2

ADDITION IN L IS COMMUTATIVE.

† There will be one exception: The Greek letter θ will be reserved for a special member of L, to be introduced in Axiom 3 below.

AXIOM 3

THERE EXISTS IN L AN ELEMENT, CALLED THE **null element** OF L, AND DENOTED θ, SUCH THAT $x + \theta = x$ FOR EVERY $x \in L$. (The $+$ in $x + \theta$ refers, of course, to addition in L.)

Note that we are entitled to refer to θ as **the** null element, because there cannot be more than one element in L satisfying the property described in Axiom 3. For a proof of this statement, see the discussion of the corresponding Axiom 3 in the previous chapter.

AXIOM 4

FOR EACH $x \in L$, THERE EXISTS AN ELEMENT $-x \in L$, CALLED THE **negative** OF x, SUCH THAT $x + (-x) = \theta$.

Once again, the definite article in "**the** negative of x" is justified in view of the uniqueness of $-x$. The proof, again, may be found in the previous chapter.

Before continuing with the axioms, let us define in L a binary operation, called **subtraction** and denoted $-$, by

$$x - y \underset{\text{def}}{=} x + (-y) \quad \text{for every} \quad x \in L, y \in L.$$

33

The field F comes into the picture at this point. A function, called **scalar multiplication** and denoted $\cdot$, is assumed to be defined on the Cartesian product $F \times L$ to L, in such a way that the following further axioms are satisfied:

AXIOM 5

FOR EVERY $\alpha \in F$, $x \in L$, and $y \in L$, THE EQUATION

$$\alpha \cdot (x + y) = \alpha \cdot x + \alpha \cdot y$$

IS SATISFIED.

Even though $\cdot$ is not a binary operation on L, it is common to say that Axiom 5 requires $\cdot$ to be *distributive* over $+$.

AXIOM 6

FOR EVERY $\alpha \in F$, $\beta \in F$, AND $x \in L$, THE EQUATION

$$(\alpha + \beta) \cdot x = \alpha \cdot x + \beta \cdot x$$

IS SATISFIED.

Note that the $+$ on the left-hand side of the equation refers to addition in F, whereas the $+$ on the right-hand side refers to addition in L. Axiom 6 resembles a distributive law, but it is not referred to as such, because of the difference in the meaning of the symbol $+$ on the two sides of the equation.

AXIOM 7

FOR EVERY $\alpha \in F$, $\beta \in F$, AND $x \in L$, THE EQUATION

$$\alpha \cdot (\beta \cdot x) = (\alpha \cdot \beta) \cdot x$$

IS SATISFIED.

The symbol $\cdot$ appears in this equation four times. In three of them it stands for *scalar multiplication* (on $F \times L$) and in one of them it stands for *multiplication* (on $F \times F$).

AXIOM 8

FOR EVERY $x \in L$, THE EQUATION

$$1 \cdot x = x$$

IS SATISFIED.

Here the symbol · is, of course, scalar multiplication, and 1 is the element of F whose existence is postulated in Axiom 7 of the previous chapter.

Given the set L and the field F, if addition and scalar multiplication are defined so as to satisfy Axioms 1–8, then we say that L is a **linear space** (or a **vector space) over the field** F. Elements of a linear space are called **vectors.**

Before looking at some examples of linear spaces, let us agree, for typographical convenience, to omit the symbol · from expressions involving scalar multiplication or multiplication in the field F. For example, the expressions $\alpha \cdot x$ and $\alpha \cdot \beta$ will henceforth be written αx and $\alpha \beta$, respectively.

EXAMPLES OF LINEAR SPACES

1. Consider a field F. Then, F is a linear space over itself. Addition in this linear space is defined as addition in F, and scalar multiplication is assumed to coincide with multiplication in F. In order to prove that F is now a linear space, we must verify that the foregoing axioms are satisfied. Axioms 1–4 are satisfied, because the very same axioms appear in the definition of a field. (See Chapter 4.) Axioms 5 and 6 above follow immediately from Axioms 6 and 9 in Chapter 4. Axiom 7 above follows from Axiom 5 in Chapter 4, and Axiom 8 follows from Axioms 6 and 7 in Chapter 4. So, F is indeed a field, with 0 playing the role of the null element θ.

2. Let F be a field, and consider the set $\{0\}$, whose only element is the zero of F. Then, the set $\{0\}$ is a (rather trivial) linear space over F, with addition defined to coincide with addition in F and scalar multiplication defined to coincide with multiplication in F. That the axioms of a linear space are satisfied is already known from the previous example. However, in order to conclude that $\{0\}$ is a linear space, we have to verify that addition and scalar multiplication are in fact functions with ranges of values in $\{0\}$. In other words, we must verify that performing additions and (scalar) multiplications on elements of $\{0\}$ will not take us outside the set $\{0\}$. But this is obviously so, since $0 + 0 = 0$ and $\alpha 0 = 0$ for every $\alpha \in F$.

3. Consider a nonempty set S and a field F. Let $\mathscr{U}$ be the family of all functions on S to F. Then, $\mathscr{U}$ is a linear space over F if we define addition and scalar multiplication as follows:

$$(f + g)(s) \underset{\text{def}}{=} f(s) + g(s) \qquad \text{for every} \quad f \in \mathscr{U}, g \in \mathscr{U}, s \in S.$$
$$(\alpha f)(s) \quad \underset{\text{def}}{=} \alpha f(s) \qquad \text{for every} \quad \alpha \in F, f \in \mathscr{U}, s \in S.$$

The fact that, with these definitions, $\mathscr{U}$ is a linear space over F (i.e., Axioms 1–8 are satisfied) is left as an exercise to the reader. Note that the null element θ of $\mathscr{U}$ is given by the function f on S to F satisfying the condition

$$f(s) = 0 \quad \text{for every} \quad s \in S.$$

A linear space over the real number system **R** is referred to as **a real linear space** (or as a **real vector space**). In this special case, one usually refers to scalar multiplication simply as **multiplication by a real number**. Throughout the remainder of this volume, we shall study various traits of real linear spaces. We concentrate on *real* linear spaces primarily as a matter of concreteness. Most of the discussion in the sequel applies equally well to linear spaces over other fields. In other words, in many of the propositions below, the hypothesis that the linear space in question is a *real* linear space is entirely inessential to the argument. Should the reader prefer generality to concreteness, he will have no difficulty converting these propositions on real linear spaces into propositions on arbitrary linear spaces.

Speaking of propositions, let us from now on assign a number to each of them, in order to facilitate cross-reference. Propositions will be numbered by chapter and consecutively within each chapter.

Proposition 5.1

Let L be real linear space. Then, the following statements hold:
   (i) *For every $x \in L$, $y \in L$, and $z \in L$, $x + y = x + z$ implies $y = z$.* (In other words, addition satisfies the law of cancellation.)
   (ii) *For every $x \in L$, $y \in L$, and $\alpha \in$ **R**, $\alpha x = \alpha y$ implies $x = y$, provided $\alpha \neq 0$.*
   (iii) *For every $x \in L$, $0x = \theta$.*
   (iv) *For every $\alpha \in$ **R**, $\alpha\theta = \theta$.*
   (v) *For every $x \in L$, $(-1)x = -x$.*

PROOF:

Left to the reader. (In problems 4.1 and 4.2 at the end of the previous chapter, similar assertions were to be proved, except that there we were dealing with a field and not with a linear space.) ∎

Consider a real linear space L, and let M be a subset of L. We say that M is **closed under addition** iff $(x + y) \in M$ whenever $x \in M$ and $y \in M$. Similarly, we say that M is **closed under multiplication by a real number** iff $(\alpha x) \in M$ whenever $x \in M$ and $\alpha \in$ **R**. Note that if L is a real linear space and if M is a subset of L that is closed both under addition and under multiplication by a real number, then M is itself a real linear space. Addition and multiplication by a real number satisfy Axioms 1–8 in M by virtue of their satisfying these axioms in L. This leads us to the following definition: A subset of a real linear space L, which is closed under addition and under multiplication by a real number, is referred to as a **linear subspace** of L. For example, if L is a real linear space, then the set $\{\theta\}$, i.e., the singleton consisting of the null element of L alone, is a linear subspace of L. Other examples of linear subspaces will arise following introduction of the notion of a linear transformation.

Let L and M be two real linear spaces. A function T on L to M is said to be a **linear transformation** (or a **linear operator**) iff the following two conditions hold:

(i) $T(x + y) = T(x) + T(y)$ for every $x \in L$ and $y \in L$;

(ii) $T(\alpha x) = \alpha T(x)$ for every $x \in L$ and $\alpha \in \mathbf{R}$.

Note that in the left-hand sides of these conditions we have addition and multiplication by a real number as defined in L, while in the right-hand sides we have the same operations as defined in M.

Conditions (i) and (ii) above are the translation into symbols of the assertion that the function T preserves sums and products by real numbers. Thus, a linear transformation is a homomorphism. We have defined a homomorphism (in Chapter 3) as a "structure-preserving" function, and a linear space may be looked upon as a set structured in such a way that certain elements are sums or scalar products of other elements.

We turn now to some properties of linear transformations.

Proposition 5.2

Let T be a linear transformation on a real linear space L to a real linear space M. Then $\mathscr{R}(T)$, the range of T, is a linear subspace of M.

PROOF:

We must show that $\mathscr{R}(T)$ is closed under addition and under multiplication by a real number. Let u and v be two arbitrary members of $\mathscr{R}(T)$. By the definition of $\mathscr{R}(T)$, we have $u = T(x)$ and $v = T(y)$ for some $x \in L$ and $y \in L$. But T is a linear transformation, and therefore $T(x + y) = T(x) + T(y) = u + v$. So, $u + v$ is the image of $x + y$ under T, so that $(u + v) \in \mathscr{R}(T)$. In a similar way, one shows that if $u \in \mathscr{R}(T)$, then $(\alpha u) \in \mathscr{R}(T)$ for every $\alpha \in \mathbf{R}$. ∎

There is another linear subspace associated with each linear transformation. This time, it is a subspace of the *domain* of the transformation. Let L and M be real linear spaces, and let T be a linear transformation on L to M. Define the set $\mathscr{N}(T)$ as follows:

$$\mathscr{N}(T) = \{x \mid x \in L \quad \text{and} \quad T(x) = \theta_M\}$$

where θ_M stands for the null element of M.

Proposition 5.3

$\mathcal{N}(T)$ is a linear subspace of L.

PROOF:

Consider two elements of $\mathcal{N}(T)$, say x and y. Then, $T(x + y) = T(x) + T(y) = \theta_M$. The first equality follows from the fact that T is a linear transformation, and the second equality follows from x and y being in $\mathcal{N}(T)$ and from the fact that $\theta_M + \theta_M = \theta_M$. Thus, $(x + y) \in \mathcal{N}(T)$, and a similar argument shows that if $x \in \mathcal{N}(T)$ and $\alpha \in \mathbf{R}$ then $(\alpha x) \in \mathcal{N}(T)$. ∎

The subspace $\mathcal{N}(T)$ is called the **null space** of the linear transformation T. Note that if T is a linear transformation on L to M, then θ_L, the null element of L, is always in $\mathcal{N}(T)$. In some cases, θ_L is the *only* member of $\mathcal{N}(T)$, and then we say that T has a **trivial null space**. The following proposition shows the importance of these notions.

Proposition 5.4

A linear transformation has a trivial null space iff it is one-to-one.

PROOF:

Let L and M be real linear spaces, and consider a linear transformation T on L to M. Suppose that $\mathcal{N}(T)$, the null space of T, is not trivial. This means that there exists an $x \in L$ such that $x \neq \theta_L$ and $T(x) = \theta_M$. We know, however, that $T(\theta_L) = \theta_M$, and so we have two distinct elements of L having the same image under T. In other words, T is not one-to-one. Conversely, suppose that T is not one-to-one. Then, there exist two elements of L, say x and y, such that $x \neq y$ and $T(x) = T(y)$. By the fact that T is a linear transformation, we have $T(x - y) = T(x) - T(y) = \theta_M$. But $x - y \neq \theta_L$. Hence, $\mathcal{N}(T)$ is not trivial. ∎

This proposition tells us that a linear transformation is invertible iff it is onto and has a trivial null space. Let L and M be real linear spaces, and let T be a linear transformation on L to M. By Proposition 5.2, the range of T, $\mathcal{R}(T)$, is a subspace of M, and so we may look upon T as a linear transformation on L onto $\mathcal{R}(T)$. When looked upon in this way, T is invertible iff it has a trivial null space.

In the next two propositions, we assert that the inverse of a linear transformation (if it exists) and the composition of two linear transformations are again linear transformations.

Proposition 5.5

Let L and M be real linear spaces, and let T be an invertible linear transfor-
mation on L onto M. Then, T^{-1} is a linear transformation on M onto L.

PROOF:

We have to show that T^{-1} preserves sums and products by real numbers. Let u and v be arbitrary members of M, and let $T^{-1}(u)$ and $T^{-1}(v)$
be denoted x and y, respectively. By the definition of the inverse function, we have $u = T(x)$ and $v = T(y)$. Now T is a linear transformation,
and therefore $T(x + y) = T(x) + T(y) = u + v$, and it follows immediately from this equation that $T^{-1}(u + v) = x + y = T^{-1}(u) + T^{-1}(v)$.
So, T^{-1} preserves sums. In a similar manner one shows that $T^{-1}(\alpha u)$
$= \alpha T^{-1}(u)$. ∎

Proposition 5.6

Let L, M, and N be real linear spaces, and let T_1 and T_2 be linear transforma-
tions on L to M and on M to N, respectively. Then, the composition $T_2 \circ T_1$ is
a linear transformation on L to N.

PROOF:

Left to the reader. ∎

Consider two real linear spaces, L and M. An invertible linear transformation on L onto M is called an **isomorphism**.

Proposition 5.7

Let L and M be two real linear spaces, and write $L \equiv M$ iff there exists an
isomorphism on L onto M. Then $\equiv$ is an equivalence relation between real
linear spaces. More precisely, if we let $\mathscr{L}$ denote the set of all real linear spaces,
then $\equiv$ is an equivalence relation in $\mathscr{L}$.

PROOF:

(i) Reflexivity: The identity function on L is certainly an isomorphism
on L onto L. In other words, the assertion $L \equiv L$ holds for every $L \in \mathscr{L}$.
(ii) Symmetry: Suppose that $L \equiv M$. This means that there exists an
isomorphism, say T, on L onto M. By Proposition 5.5 and by the
properties of invertible functions, T^{-1} must be an isomorphism on M
onto L, and therefore $M \equiv L$.
(iii) Transitivity: Suppose that $L \equiv M$ and $M \equiv N$. By Proposition
5.6 and by the properties of invertible functions, one obtains immediately
that $L \equiv N$. ∎

If L and M are real linear spaces and there exists an isomorphism on L onto M, then one says that L and M are **isomorphic**. We know that every equivalence relation can be used to partition its domain into a family of disjoint equivalence classes. Here, an equivalence class is simply the set of all real linear spaces that are isomorphic to a given real linear space. In the study of linear spaces, it suffices to examine just one representative of each equivalence class, and this accounts for the importance of the concept of isomorphism. What we claim here is that the study of a real linear space constitutes a study of all the real linear spaces that are isomorphic to it. Consider two isomorphic real linear spaces, L and M, and let T be an isomorphism on L onto M. Suppose also that $\mathscr{A}_L$ is an assertion of some sort about the space L. Now let us modify the assertion $\mathscr{A}_L$ in the following manner: (i) If a member x of L appears in $\mathscr{A}_L$, replace it by $T(x)$. (ii) If the symbols $+$ and $\cdot$ appear in $\mathscr{A}_L$, leave them unchanged, but interpret them as addition and multiplication by a real number in M. (iii) Leave all logical connectives and quantifiers unchanged. It can be shown that as a result of all this, the assertion $\mathscr{A}_L$ converts into an assertion $\mathscr{A}_M$ about the space M and that, furthermore, $\mathscr{A}_L$ is true iff $\mathscr{A}_M$ is true. Thus, studying the space L also constitutes a study of the space M or of any other space that is isomorphic to L.

5.1. Consider the Cartesian product $\mathbf{R} \times \mathbf{R}$, i.e., the set of all pairs of real numbers. Define addition in this Cartesian product by

$$\langle \alpha, \beta \rangle + \langle \alpha', \beta' \rangle \underset{\text{def}}{=} \langle \alpha + \alpha', \beta + \beta' \rangle$$

where α, β, α', and β' are in $\mathbf{R}$ and the $+$ on the right-hand side refers to addition in $\mathbf{R}$. Now define multiplication by a real number by

$$\gamma \langle \alpha, \beta \rangle \underset{\text{def}}{=} \langle \gamma\alpha, \gamma\beta \rangle$$

where α, β, and γ are in $\mathbf{R}$ and the products on the right-hand side refer to multiplication in $\mathbf{R}$. Show that with these definitions, $\mathbf{R} \times \mathbf{R}$ becomes a real linear space. (This space will be denoted $\mathbf{R}^2$.)

5.2. In Problem 5.1, change the definition of addition as follows:

$$\langle \alpha, \beta \rangle + \langle \alpha', \beta' \rangle \underset{\text{def}}{=} \langle \alpha + \alpha', 0 \rangle.$$

With this modification, is $\mathbf{R} \times \mathbf{R}$ a real linear space? Prove your answer.

5.3. In Problem 5.1, change the definition of multiplication by a real number as follows:

$$\gamma\langle\alpha, \beta\rangle \underset{\text{def}}{=} \langle\gamma\alpha, \beta\rangle.$$

With this modification, is $\mathbf{R} \times \mathbf{R}$ a real linear space? Prove your answer.

5.4. For each real number α, define the **integer part** of α, denoted $[\alpha]$, by the conditions

(i) $[\alpha]$ is an integer;

(ii) $[\alpha] \leq \alpha$;

(iii) $[\alpha] + 1 > \alpha$.

Also, define the **pure fraction** part of α, denoted $]\alpha[$, by

$$]\alpha[= \alpha - [\alpha].$$

Finally, define the **semi-open unit interval**, U, by

$$U = \{\alpha \,|\, \alpha \in \mathbf{R}, 0 \leq \alpha < 1\}.$$

Addition and multiplication by a real number are now defined in U by

$$x + y \underset{\text{def}}{=}]x + y[\quad \text{for every} \quad x \in U, y \in U,$$
$$\alpha x \underset{\text{def}}{=}]\alpha x[\qquad \text{for every} \quad x \in U, \alpha \in \mathbf{R}.$$

(In both definitions, the right-hand side involves the usual arithmetic operations in $\mathbf{R}$.) With these definitions, is U a real linear space? Prove your answer.

5.5. Prove the assertion that the composition of two linear transformations is a linear transformation.

5.6. Consider the space $\mathbf{R}^2$, as introduced in Problem 5.1. The following equations determine five functions (all denoted f) on $\mathbf{R}^2$ to $\mathbf{R}^2$:

(i) $f(\langle\alpha, \beta\rangle) = \langle\alpha + \beta, \alpha + \beta\rangle$ for every $\langle\alpha, \beta\rangle \in \mathbf{R}^2$;

(ii) $f(\langle\alpha, \beta\rangle) = \langle\alpha\beta, \alpha\beta\rangle$ for every $\langle\alpha, \beta\rangle \in \mathbf{R}^2$;

(iii) $f(\langle\alpha, \beta\rangle) = \langle\beta, 0\rangle$ for every $\langle\alpha, \beta\rangle \in \mathbf{R}^2$;

(iv) $f(\langle\alpha, \beta\rangle) = \langle\alpha^2, \beta^2\rangle$ for every $\langle\alpha, \beta\rangle \in \mathbf{R}^2$;

(v) $\begin{aligned} f(\langle\alpha, \beta\rangle) &= \langle\alpha, \alpha\rangle \quad \text{if} \quad \alpha \leq \beta \\ &= \langle\beta, \beta\rangle \quad \text{if} \quad \beta \leq \alpha \end{aligned}$ for every $\langle\alpha, \beta\rangle \in \mathbf{R}^2$.

Note that the expression $f(\langle\alpha, \beta\rangle)$ would usually be written simply as $f(\alpha, \beta)$.

However, it seems that in the present context it is best to leave the pointed brackets in. Now to the questions: Which of these functions f is a linear transformation? Prove your assertions. In cases where f is in fact a linear transformation, find its range and its null space. Finally, you may be familiar with the fact that members of $\mathbf{R}^2$ can be represented diagrammatically as points in the plane, with a system of Cartesian coordinates. For example, the point $\langle \alpha, \beta \rangle$ is represented by measuring α units on the horizontal axis and β units on the vertical axis, dropping perpendiculars, and marking the point of intersection. Returning now to the problem at hand, in each case where f is a linear transformation, draw a diagram, displaying the range and the null space of f.

5.7. Let L be a real linear space, and let M and N be linear subspaces of L. Is $M \cup N$ a subspace of L? Is $M \cap N$ a subspace of L? Prove your assertions.

6   Finite-Dimensional

Spaces

Consider a real linear space L, and let $S = \{x_1, x_2, \ldots, x_n\}$ be a finite subset of L. Furthermore, let x be a member of L. If x is of the form

$$x = \alpha_1 x_1 + \alpha_2 x_2 + \cdots + \alpha_n x_n$$

for some real numbers $\alpha_1, \alpha_2, \ldots, \alpha_n$, then x is said to be a **linear combination** of $x_1, x_2, \ldots, x_n$.

As a matter of notation, it is customary to use the symbol $\sum$ to designate a sum. Specifically, the above equation, expressing x in terms of $x_1, x_2, \ldots, x_n$, would in many cases be written as follows:

$$x = \sum_{i=1}^{n} \alpha_i x_i.$$

Note that $\sum$ is used to denote summation according to the definition of addition in any given linear space.

Let us return now to the real linear space L and to its subset S, as introduced in the first paragraph above. Consider the set

$$\left\{ x \mid x = \sum_{i=1}^{n} \alpha_i x_i \quad \text{for some real numbers} \quad \alpha_1, \alpha_2, \ldots, \alpha_n \right\},$$

i.e., the set of all possible linear combinations of the members of S. This set is clearly a subset of L and it is referred to as the subset of L that is **spanned** (or **generated**) by S.

Proposition 6.1

Let $S = \{x_1, x_2, \ldots, x_n\}$ be a subset of a real linear space L. Then, the set spanned by S is a linear subspace of L.

PROOF:

Let the set spanned by S be denoted P, and let x and y be arbitrary members of P. Then, by definition, $x = \sum \alpha_i x_i$ for some real numbers $\alpha_1, \alpha_2, \ldots, \alpha_n$ and $y = \sum \beta_i x_i$ for some real numbers $\beta_1, \beta_2, \ldots, \beta_n$. This leads to

$$x + y = \sum \alpha_i x_i + \sum \beta_i x_i$$
$$= \sum (\alpha_i + \beta_i) x_i,$$

by Axioms 1, 2, and 6 in the definition of a linear space. Hence, $x + y$ is also an element of P. Using Axiom 7 in the definition of a linear space, one can show also that if $x \in P$ then $(\alpha x) \in P$ for every real number α. ∎

45

A subset $S = \{x_1, x_2, \ldots, x_n\}$ of a real linear space is said to be **linearly independent** iff the equation

$$\sum_{i=1}^{n} \alpha_i x_i = \theta$$

implies that $\alpha_i = 0$ for $i = 1, 2, \ldots, n$. In other words, S is a linearly independent set iff it is impossible to find n real numbers, $\alpha_1, \alpha_2, \ldots, \alpha_n$, not all zero, such that $\sum \alpha_i x_i = \theta$. A set that is not linearly independent is said to be **linearly dependent**. It is also common to refer to the *members* of a linearly independent set as being linearly independent. For example, the statement "x and y are linearly independent" means that the set $\{x, y\}$ is a linearly independent set.

Proposition 6.2

Let $S = \{x_1, x_2, \ldots, x_n\}$ be a linearly independent subset of a real linear space. Then, the following two assertions are true:
  (i) *$\theta \notin S$;*
  (ii) *Every subset of S is also a linearly independent set.*

PROOF:
Left to the reader. ∎

Proposition 6.3

Once again, let $S = \{x_1, x_2, \ldots, x_n\}$ be a subset of a real linear space. S is linearly independent iff no member of S is a linear combination of the other members of S.

PROOF:
Assume that some member of S is a linear combination of the other members. Without loss of generality, we can take this member to be x_1. Thus,

$$x_1 = \sum_{i=2}^{n} \alpha_i x_i$$

for some real numbers, $\alpha_2, \ldots, \alpha_n$. Define $\alpha_1 = -1$. Clearly, we now have

$$\sum_{i=1}^{n} \alpha_i x_i = \theta$$

and the α's are not all zero, so that S is linearly dependent. Conversely, if S is linearly dependent, then there exist real numbers $\alpha_1, \alpha_2, \ldots, \alpha_n$, not all zero, such that $\sum \alpha_i x_i = \theta$. Again without loss of generality, we may assume that $\alpha_1 \neq 0$. Then,

$$x_1 = \sum_{i=2}^{n} (-\alpha_i/\alpha_1) x_i,$$

so that x_1 is a linear combination of the other members of S. ∎

We come now to the most important proposition of this chapter. Its importance lies not only in its content, but also in the method used to prove it. This method may be used as a computational procedure in several applications.

Proposition 6.4

Let $S = \{x_1, x_2, \ldots, x_n\}$ *and* $Q = \{y_1, y_2, \ldots, y_m\}$ *be two subsets of a real linear space* L. (Note that S has n elements and Q has m elements.) *If* S *spans the space* L *and* Q *is linearly independent, then* $n \geq m$.

PROOF:
In order to establish this assertion, we need to perform a sequence of steps.

Step 1: For the sake of symmetry with the following steps, rename the set S and call it S_1. Now consider the element y_1 of Q. If $y_1 \in S_1$, then rename S_1, call it S_2, and go to Step 2. If $y_1 \notin S_1$, then consider the set obtained by adjoining y_1 to S_1, i.e., the set $\{y_1, x_1, x_2, \ldots, x_n\}$. Since S_1 spans L, y_1 is a linear combination of the x's. In other words, there exist real numbers $\alpha_1, \alpha_2, \ldots, \alpha_n$ such that

$$y_1 = \sum_{i=1}^{n} \alpha_i x_i.$$

Note that these α's cannot all be zero. For if they were all zero, we would have had

$$y_1 = \sum 0 x_i = \theta.$$

But y_1 is a member of a linearly independent set, and therefore (Proposition 6.2) it cannot be the null element. Thus, at least one of the α's, say α_j, is not zero. This means that x_j must be a linear combination of the other x's and y_1. To see this, recall that $y_1 = \sum \alpha_i x_i$ and, by isolating x_j in this equation, one obtains

$$x_j = (1/\alpha_j) y_1 + \sum_{i \neq j} (-\alpha_i/\alpha_j) x_i,$$

which shows x_j as a linear combination of the other x's and y_1, provided $\alpha_j \neq 0$. Now consider the set

$$\{y_1, x_1, x_2, \ldots, x_{j-1}, x_{j+1}, \ldots, x_n\},$$

i.e., the set S_1 with x_j thrown out and y_1 brought in. Call this set S_2. We claim that S_2 spans the space L. To see this, consider an arbitrary element z of L. Since, by hypothesis, the set S_1 spans the space L, there exist real numbers $\gamma_1, \gamma_2, \ldots, \gamma_n$ such that

$$z = \sum_{i=1}^{n} \gamma_i x_i.$$

But now, if we substitute for x_j in this equation its expression in terms of the other x's and y_1 (see above), we find that z is a linear combination of the elements of S_2. In other words, S_2 spans the space L.

Step 2: Consider the element y_2 of Q. If $y_2 \in S_2$, then rename the set S_2, call it S_3, and go to Step 3. If $y_2 \notin S_2$, then adjoin y_2 to the set S_2, to obtain the set

$$\{y_2, y_1, x_1, x_2, \ldots, x_{j-1}, x_{j+1}, \ldots, x_n\}.$$

Since S_2 spans L, we can express y_2 as a linear combination of the members of S_2, i.e., we can find real numbers $\beta_1, \alpha_1, \alpha_2, \ldots, \alpha_{j-1}, \alpha_{j+1}, \ldots, \alpha_n$ such that

$$y_2 = \beta_1 y_1 + \sum_{i \neq j} \alpha_i x_i.$$

We claim that the α's in this last equation cannot all be zero. (Note that these are not the α's that appeared in Step 1.) If all the α's in the foregoing equation were zero, we would have

$$y_2 = \beta_1 y_1,$$

which contradicts the assumption that $\{y_1, y_2\}$, being a subset of a linearly independent set, is itself linearly independent. So, suppose $\alpha_k \neq 0$. If we now remove x_k from the set and retain y_2, we shall once again obtain a set that spans L. The proof of this assertion is precisely as in Step 1. Let us call our new set (with x_k removed and y_2 retained) S_3, and proceed to the next step.

Step 3: If $y_3 \in S_3$, rename S_3, call it S_4, and go to Step 4. If $y_3 \notin S_3$, bring y_3 into S_3 and write

$$y_3 = \beta_1 y_1 + \beta_2 y_2 + \sum_{\substack{i \neq j \\ i \neq k}} \alpha_i x_i.$$

At least one of the α's, say α_h, is not zero. This follows from the assumption that $\{y_1, y_2, y_3\}$ is a subset of a linearly independent set and is therefore itself a linearly independent set. Now remove the element x_h and call the set that emerges S_4. Note that S_4 spans the space L, and go to the next step.

We shall have the proof of our proposition after answering the following question: How many steps of the kind just described can we take before the procedure comes to a halt? Clearly, the procedure cannot go on forever, because sooner or later either the set S or the set Q will be exhausted. In the former case, we shall have to stop because there will be no more x's to throw out, and in the latter case we shall have to stop because there will be no more y's to bring in. It remains to be noted that *the set S cannot be exhausted before the set Q*. For suppose S were exhausted before Q. This would mean that after n steps we would arrive at a set S_{n+1}, which is a *proper subset* of Q and which spans L. But this means that the remaining elements of Q (i.e., the elements of $Q \sim S_{n+1}$) are linear combinations of elements in a proper subset of Q, and this contradicts the hypothesis that Q is a linearly independent set. Hence, S must have at least as many elements as Q, i.e., $n \geq m$. ∎

The process of removing an element from the set S and introducing in its stead an element from Q, as described in the foregoing proof, is sometimes called a **pivot**. There are many problems in which performing a well-chosen sequence of pivots constitutes a key to the solution. A prominent case in point is *linear programming*. In a linear programming problem, one is given a finite set of elements in a linear space, and the task is to find a subset of this original set that has a certain optimality property. It is, of course, possible to find this optimal subset by using an exhaustive search (since the number of candidate subsets is finite), but this is very inefficient. The best known *efficient* method for finding the solution to a linear programming problem is known as the *simplex method*. This method rests on the notion of a pivot: One starts out with a provisional subset, which is then improved upon gradually via a sequence of pivots, until the optimal subset is reached. A full-scale exposition of linear programming goes beyond the scope of this volume, so let us return now to the main line of the discussion.

Let L be a real linear space. A linearly independent subset B of L that spans L is called a **linear basis**, or simply a **basis**, for L. Note that if B is a basis for L, then no proper subset of B is a basis for L. (Why?) A basis, then, is a set that can be used, by means of linear combinations, to generate the entire space L and, furthermore, a set that is *efficient* in the sense that no proper subset of it will generate the entire space L. (For this reason, a basis is sometimes referred to as a **minimal generating subset** of a linear space.) A real linear space for which there exists a finite basis is said to be **finite-dimensional** (or **finitely generated**).

For convenience, let us henceforth use the term "finite-dimensional space" to mean "finite-dimensional real linear space."

Corollary 1 to Proposition 6.4

Let L be a finite-dimensional space. Then, all bases for L have the same number of elements.

PROOF:

Let $B_1 = \{x_1, x_2, \ldots, x_n\}$ and $B_2 = \{y_1, y_2, \ldots, y_m\}$ be two bases for L. Apply Proposition 6.4 twice, once with B_1 as S and B_2 as Q, and once with B_2 as S and B_1 as Q. The result will be $n \geq m$ and $m \geq n$. Hence, we must have $m = n$. ∎

Let L be a finite-dimensional space, and suppose that a basis for L has n elements. Then, n is referred to as the **dimension** of L, and one writes

$$dim \, L = n.$$

Furthermore, L in this case is said to be an **n-dimensional** space. If L has only one element (i.e., the null element), then L is defined to be zero-dimensional. This is consistent with the fact that in a trivial linear space, a linearly independent set is necessarily empty.

Corollary 2 to Proposition 6.4

A linearly independent set in an n-dimensional space cannot have more than n elements.

PROOF:
This is merely a restatement of Proposition 6.4. ∎

Proposition 6.5

Let L_n be an n-dimensional (real linear) space. Every linearly independent set in L_n is a subset of a basis for L_n, and every n-element linearly independent set in L_n is a basis for L_n.

PROOF:

Let $B = \{x_1, x_2, \ldots, x_n\}$ be a basis for L_n and let $A = \{y_1, y_2, \ldots, y_m\}$ be a linearly independent subset of L_n. We shall establish the desired result by showing that performing a pivot on the set B preserves the linear independence of B. (We already know, from Proposition 6.4, that performing a pivot preserves the spanning property of B.) So, let us take the element y_1 of A and adjoin it to the basis B. Since B spans L_n, there exist real numbers $\alpha_1, \alpha_2, \ldots, \alpha_n$ such that

$$y_1 = \sum_{i=1}^{n} \alpha_i x_i.$$

The α's cannot all be zero (see Proposition 6.4), so we can find an x_j for which $\alpha_j \neq 0$, and throw this x_j out. As a result, we obtain the following postpivot set:

$$\{y_1, x_1, x_2, \ldots, x_{j-1}, x_{j+1}, \ldots, x_n\}.$$

We must show that this set is still linearly independent. Suppose it is not. Then, there exist real numbers $\gamma_1, \beta_1, \beta_2, \ldots, \beta_{j-1}, \beta_{j+1}, \ldots, \beta_n$, *not all zero*, such that

$$\gamma_1 y_1 + \sum_{i \neq j} \beta_i x_i = \theta.$$

Using the equation for y_1 above, we get, following rearrangement,

$$\gamma_1 \alpha_j x_j + \sum_{i \neq j} (\gamma_1 \alpha_i + \beta_i) x_i = \theta.$$

In other words, we have found real numbers $\delta_1, \delta_2, \ldots, \delta_n$ such that

$$\sum_{i=1}^{n} \delta_i x_i = \theta.$$

Now, it is easily seen that the δ_i's cannot all be zero. (Recall that $\alpha_j \neq 0$.) Therefore, this last equation contradicts the linear independence of B. Thus, pivoting preserves linear independence, as well as the spanning property. The proposition is now proved by noting that we can introduce all the elements of A into a basis for L_n by performing a sequence of m pivots. ∎

Proposition 6.6

Let L_n be an n-dimensional space, and let M be a linear subspace of L_n. Then, M is finite-dimensional, and dim $M \leq n$.

PROOF:

Pick a nonnull element of M, say x_1. (If M has no nonnull elements, i.e., if $M = \{\theta\}$, then the proposition is trivial.) If the set $\{x_1\}$ spans M, then the proposition is proved, because M is 1-dimensional and $1 \leq n$ by the second corollary to Proposition 6.4. Suppose, therefore, that $\{x_1\}$ does not span M. Then, there exists in M an element, say x_2, such that $\{x_1, x_2\}$ is a linearly independent set. If $\{x_1, x_2\}$ spans M, the proposition is proved. If $\{x_1, x_2\}$ does not span M, then there exists in M an element x_3 such that $\{x_1, x_2, x_3\}$ is a linearly independent set. Once again, the proof is complete if $\{x_1, x_2, x_3\}$ spans M; if not, we go to the next step. All that needs to be shown now is that the procedure just described must come to a halt after no more than n steps. But this is obvious because if the procedure were to go on for more than n steps, the result would be a linearly independent subset of L_n with more than n elements, which would contradict Proposition 6.4. ∎

Let $\{x_1, x_2, \ldots, x_n\}$ be a basis for an n-dimensional space L_n. It is customary to impose on the elements of this basis some arbitrary order, so that, say, x_1 becomes the *first component* of the basis, x_2 the *second component*, and so on. By imposing such an arbitrary order, one converts the set $\{x_1, x_2, \ldots, x_n\}$ into the n-tuple $\langle x_1, x_2, \ldots, x_n \rangle$. This n-tuple is referred to as an **ordered basis** for L_n. From now on, we shall use the word "basis" to mean "ordered basis." Bold-faced Roman characters will be used to denote ordered bases.

Proposition 6.7

Let $\mathbf{B} = \langle x_1, x_2, \ldots, x_n \rangle$ *be a basis for an n-dimensional space* L_n, *and let* y *be some member of* L_n. *There exists one and only one n-tuple of real numbers* $\langle \alpha_1, \alpha_2, \ldots, \alpha_n \rangle$ *such that* $y = \sum \alpha_i x_i$.

PROOF:

That there exists one such n-tuple is merely an expression of the fact that $\mathbf{B}$ spans L_n. To see that there cannot be more than one, suppose that we have two n-tuples of real numbers, $\langle \alpha_1, \alpha_2, \ldots, \alpha_n \rangle$ and $\langle \beta_1, \beta_2, \ldots, \beta_n \rangle$, such that $y = \sum \alpha_i x_i$ and also $y = \sum \beta_i x_i$ for some $y \in L_n$. Then, clearly,

$$\sum_{i=1}^{n} (\alpha_i - \beta_i) x_i = \theta,$$

and it follows from the linear independence of the x's that $\alpha_i = \beta_i$ for $i = 1, 2, \ldots, n$. ∎

Given a basis $\mathbf{B} = \langle x_1, x_2, \ldots, x_n \rangle$ for an n-dimensional space, and given a member y of this space, the n-tuple of real numbers $\langle \alpha_1, \alpha_2, \ldots, \alpha_n \rangle$ such that $y = \sum \alpha_i x_i$ is called the n-tuple of **coordinates** of y relative to the basis $\mathbf{B}$. α_1 is called the **first coordinate** of y relative to $\mathbf{B}$, α_2 the **second coordinate** of y relative to $\mathbf{B}$, and so on.

Our last task in this chapter is to show that all n-dimensional spaces are isomorphic (to one another) and that a space that is isomorphic to an n-dimensional space is itself n-dimensional.

Proposition 6.8

Let L_n be an n-dimensional space, and let M be an arbitrary real linear space. L_n and M are isomorphic iff M is an n-dimensional space.

PROOF:

Assume first that M is an n-dimensional space. Let **B** be a basis for L_n and let **C** be a basis for M. Consider an arbitrary member of L_n, say x, and let $\langle \alpha_1, \alpha_2, \ldots, \alpha_n \rangle$ be the n-tuple of coordinates of x relative to **B**. Define $T(x)$ to be that member of M whose n-tuple of coordinates relative to **C** is also $\langle \alpha_1, \alpha_2, \ldots, \alpha_n \rangle$. T is obviously a one-to-one function on L_n onto M. It is simple to verify, furthermore, that T is a linear transformation. Conversely, assume that there exists an isomorphism T on L_n onto M, and let $\mathbf{B} = \langle x_1, x_2, \ldots, x_n \rangle$ be a basis for L_n. We claim that the n-tuple $\langle T(x_1), T(x_2), \ldots, T(x_n) \rangle$, which might be denoted $T(\mathbf{B})$ for short, is a basis for M. First, let us check linear independence. Suppose that the components of $T(\mathbf{B})$ are *not* linearly independent. Then, there exist real numbers $\alpha_1, \alpha_2, \ldots, \alpha_n$, not all zero, such that

$$\sum_{i=1}^{n} \alpha_i T(x_i) = \theta_M$$

where θ_M is the null element of M. Since T is a linear transformation, we have

$$T\left(\sum_{i=1}^{n} \alpha_i x_i \right) = \theta_M.$$

But T is invertible, and so, by Proposition 5.4, it has a trivial null space. In other words, the last equation implies

$$\sum_{i=1}^{n} \alpha_i x_i = \theta_L$$

where θ_L is the null element of L_n. This contradicts the linear independence of the x's. To see that $T(\mathbf{B})$ spans M, pick an arbitrary member of M, say y. Since T is onto, there exists an $x \in L_n$ such that $y = T(x)$. But x may be expressed as a linear combination of the components of the basis **B**,

$$x = \sum_{i=1}^{n} \beta_i x_i,$$

for some real numbers $\beta_1, \beta_2, \ldots, \beta_n$. By the properties of a linear transformation, we now have

$$y = T(x) = \sum_{i=1}^{n} \beta_i T(x_i),$$

which means that y is a linear combination of the components of $T(\mathbf{B})$. ∎

The fact that all n-dimensional spaces are isomorphic to one another permits us to pick one specific n-dimensional space and study it as a "representative" of all n-dimensional spaces.

PROBLEMS

6.1. Write down a detailed proof of Proposition 6.2.

6.2. Let L_n be an n-dimensional space. Show that the only n-dimensional subspace of L_n is L_n itself.

6.3. Let $x_1, x_2, \ldots, x_n$, and y be members of a real linear space. Assume that the set $\{x_1, \ldots, x_n\}$ is linearly independent. Is the set $\{x_1 + y, x_2 + y, \ldots, x_n + y\}$ also linearly independent?

6.4. Consider the space $\mathbf{R}^2$, as defined in Problem 5.1. Find three different bases for $\mathbf{R}^2$, and prove that they are, indeed, bases for $\mathbf{R}^2$. Let $\langle \alpha, \beta \rangle$ be an arbitrary element of $\mathbf{R}^2$. Determine the coordinates of $\langle \alpha, \beta \rangle$ relative to each of the three bases you have found.

6.5. Draw two arbitrary straight lines through the origin in the plane, and let them be denoted, say, $\mathscr{R}$ and $\mathscr{N}$. Find a linear transformation on $\mathbf{R}^2$ to $\mathbf{R}^2$, whose range is $\mathscr{R}$ and whose null space is $\mathscr{N}$. Can you characterize *all* the linear transformations on $\mathbf{R}^2$ to $\mathbf{R}^2$ that have this property? Prove your answer.

6.6. Let L be a finite-dimensional space, and let M and N be linear subspaces of L. We define $M + N$ to be the set of all members of L that can be written as the sum of a member of M and a member of N. Show (i) that $M + N$ is a subspace of L, and (ii) that if $M + N = L$ and $M \cap N = \{\theta\}$, then $dim\ L = dim\ M + dim\ N$.

7 The Space $\mathbf{R}^n$

Let L and M be two real linear spaces, and consider the Cartesian product, $L \times M$. As it stands, this Cartesian product is not itself a real linear space. However, there is a natural way to convert it into such a space. One simply defines addition and multiplication by a real number in $L \times M$ by

$$\langle x, y \rangle + \langle w, z \rangle \underset{\text{def}}{=} \langle x + w, y + z \rangle$$
$$\lambda \langle x, y \rangle \underset{\text{def}}{=} \langle \lambda x, \lambda y \rangle$$

for every $x \in L$, $w \in L$, $y \in M$, $z \in M$, and for every $\lambda \in \mathbf{R}$. (Note that on the right-hand sides of these definitions we have addition and multiplication by a real number as defined in L and in M.) It is simple to verify by checking the axioms that, with these definitions, the product $L \times M$ becomes a real linear space. Similarly, if $L^1, L^2, \ldots, L^n$ are real linear spaces, then one can convert the Cartesian product $L^1 \times L^2 \times \cdots \times L^n$ into a real linear space by writing

$$\langle x^1, \ldots, x^n \rangle + \langle z^1, \ldots, z^n \rangle \underset{\text{def}}{=} \langle x^1 + z^1, \ldots, x^n + z^n \rangle$$
$$\lambda \langle x^1, \ldots, x^n \rangle \underset{\text{def}}{=} \langle \lambda x^1, \ldots, \lambda x^n \rangle$$

for every $x^i \in L^i$, $z^i \in L^i$, $i = 1, \ldots, n$, and for every $\lambda \in \mathbf{R}$.

In the foregoing definition, let us now substitute the real number system, **R**, for each of the spaces $L^1, \ldots, L^n$. The result will be a real linear space whose elements are n-tuples of real numbers. We shall use the symbol **R**^{n} to denote this space. To repeat, **R**^{n} is the real linear space obtained if we take the set of all n-tuples of real numbers, and define addition and multiplication by a real number on this set by

$$\langle \alpha_1, \ldots, \alpha_n \rangle + \langle \beta_1, \ldots, \beta_n \rangle \underset{\text{def}}{=} \langle \alpha_1 + \beta_1, \ldots, \alpha_n + \beta_n \rangle$$
$$\lambda \langle \alpha_1, \ldots, \alpha_n \rangle \underset{\text{def}}{=} \langle \lambda \alpha_1, \ldots, \lambda \alpha_n \rangle.$$

The notation **R**^{n} is used here because the space in question is, so to speak, the "n-th power" (in the sense of Cartesian products) of **R**.

Mainly for typographical reasons, it is common to write the elements of **R**^{n} vertically rather than horizontally, and to enclose them in square brackets. Thus, the following will be a typical element of **R**^{n}:

$$\begin{bmatrix} \alpha_1 \\ \alpha_2 \\ \cdot \\ \cdot \\ \cdot \\ \alpha_n \end{bmatrix}.$$

We shall use lowercase bold-faced Roman characters ($\mathbf{x}$, $\mathbf{y}$, $\mathbf{u}$, etc.) to denote members of $\mathbf{R}^n$. It will also be convenient to adopt the following notation for the *components* of a member of $\mathbf{R}^n$:

$$\mathbf{x} = \begin{bmatrix} x_1 \\ x_2 \\ \cdot \\ \cdot \\ \cdot \\ x_n \end{bmatrix}, \mathbf{y} = \begin{bmatrix} y_1 \\ y_2 \\ \cdot \\ \cdot \\ \cdot \\ y_n \end{bmatrix}, \mathbf{u} = \begin{bmatrix} u_1 \\ u_2 \\ \cdot \\ \cdot \\ \cdot \\ u_n \end{bmatrix}, \text{etc.}$$

In other words, x_i will always stand for the i-th component of $\mathbf{x}$, u_i—for the i-th component of $\mathbf{u}$, and so forth. Now, how shall we denote the i-th component of an element of $\mathbf{R}^n$ if the notation for this element already has a subscript in it? For example, suppose that $\mathbf{x}_1$, $\mathbf{x}_2$, ..., $\mathbf{x}_k$ are elements of $\mathbf{R}^n$. How shall we denote the i-th component of $\mathbf{x}_1$, or $\mathbf{x}_2$, or $\mathbf{x}_j$? The most prevalent convention in cases like this is to use *double* subscripts. Thus, if $\mathbf{x}_j$ is an element of $\mathbf{R}^n$, then the i-th component of $\mathbf{x}_j$ will be written x_{ij}. We shall continue to use Greek letters for real numbers appearing by themselves, but for real numbers appearing as components of a vector in $\mathbf{R}^n$ we shall use Roman characters with subscripts (and with double subscripts, if necessary).

The null element of $\mathbf{R}^n$ is, obviously, the n-tuple

$$\begin{bmatrix} 0 \\ 0 \\ \cdot \\ \cdot \\ \cdot \\ 0 \end{bmatrix}$$

whose components are all zero. We shall use the symbol $\mathbf{0}$ to denote this element. (However, we shall retain the symbol θ for the null element of an arbitrary linear space.) $\mathbf{0}$ is sometimes referred to as the **origin** of $\mathbf{R}^n$.

In order to determine the dimension of the space $\mathbf{R}^n$, let us now introduce n special members of $\mathbf{R}^n$, known as the **unit vectors**. Consider the element of $\mathbf{R}^n$ that is obtained by taking the null element, $\mathbf{0}$, and changing its i-th component from 0 to 1. This element is called the **i-th unit vector** of $\mathbf{R}^n$, and it is denoted $\mathbf{e}_i$. Thus,

$$\mathbf{e}_1 = \begin{bmatrix} 1 \\ 0 \\ 0 \\ \cdot \\ \cdot \\ \cdot \\ 0 \end{bmatrix}, \mathbf{e}_2 = \begin{bmatrix} 0 \\ 1 \\ 0 \\ \cdot \\ \cdot \\ \cdot \\ 0 \end{bmatrix}, \ldots, \mathbf{e}_n = \begin{bmatrix} 0 \\ 0 \\ 0 \\ \cdot \\ \cdot \\ \cdot \\ 1 \end{bmatrix}.$$

Proposition 7.1

The n-tuple $\langle \mathbf{e}_1, \mathbf{e}_2, \ldots, \mathbf{e}_n \rangle$ of unit vectors is a basis for $\mathbf{R}^n$. Consequently, $\mathbf{R}^n$ is an n-dimensional space.

PROOF:

Let us first show linear independence. Suppose that

$$\sum_{i=1}^{n} \alpha_i \mathbf{e}_i = \mathbf{0}$$

for some real numbers $\alpha_1, \alpha_2, \ldots, \alpha_n$. We must show that $\alpha_i = 0$ for every i. By the definition of the unit vectors, $\sum \alpha_i \mathbf{e}_i$ is that element of $\mathbf{R}^n$ whose first component is α_1, whose second component is α_2, and so on. Hence, $\sum \alpha_i \mathbf{e}_i = \mathbf{0}$ does, indeed, imply that $\alpha_i = 0$ for every i. To see that the unit vectors span the space $\mathbf{R}^n$, merely observe that for every member $\mathbf{x}$ of $\mathbf{R}^n$ (with components x_i) we have

$$\mathbf{x} = \sum_{i=1}^{n} x_i \mathbf{e}_i.$$

This completes the proof. ∎

The *n*-tuple of unit vectors $\langle \mathbf{e}_1, \mathbf{e}_2, \ldots, \mathbf{e}_n \rangle$ is called the **unit basis** for $\mathbf{R}^n$, and it is denoted **I**. We have seen in the foregoing proof that the *n*-tuple of coordinates of a member $\mathbf{x}$ of $\mathbf{R}^n$ relative to the unit basis is simply $\mathbf{x}$ itself. (This is precisely what the last equation in the proof above says.) It is this fact that makes working with the unit basis so convenient.

The fact that $\mathbf{R}^n$ is an *n*-dimensional space means, of course, that it is isomorphic to any other *n*-dimensional space. A very natural isomorphism between an arbitrary *n*-dimensional space and $\mathbf{R}^n$ may be constructed as follows: Let L_n be an *n*-dimensional space, and let **B** be a basis for L_n. For each member x of L_n, let $T(x)$ be the *n*-tuple of coordinates of x relative to **B**. Then, T is an isomorphism on L_n onto $\mathbf{R}^n$. (We leave the proof of this assertion to the reader.) In other words, by substituting for a member of L_n its n-tuple of coordinates (relative to some fixed basis) we are able to reduce the study of L_n to a study of $\mathbf{R}^n$. For this reason, people sometimes use the term "n-dimensional space" and the term "the space $\mathbf{R}^n$" as though they were synonyms.

We shall now proceed to define in $\mathbf{R}^n$ a new operation that, in principle, need not be defined in other n-dimensional spaces. There is a way to characterize this new operation in terms of linear transformations on $\mathbf{R}^n$ to $\mathbf{R}$ (see problem 7.3 at the end of this chapter), but rather than take this roundabout route, we shall define the operation directly. Let $\mathbf{x}$ (with components x_i) and $\mathbf{y}$ (with components y_i) be two members of $\mathbf{R}^n$. The real number

$$\sum_{i=1}^{n} x_i y_i$$

is called the **inner product** of $\mathbf{x}$ and $\mathbf{y}$, and is denoted $(\mathbf{x}, \mathbf{y})$. The function that associates the real number $(\mathbf{x}, \mathbf{y})$ with the pair $\langle \mathbf{x}, \mathbf{y} \rangle$ is called the **inner product function**. It is, of course, a function on $\mathbf{R}^n \times \mathbf{R}^n$ onto $\mathbf{R}$.

The inner product $(\mathbf{x}, \mathbf{y})$ may also be written $\mathbf{x}'\mathbf{y}$. This alternative notation arises from certain operations on matrices, which we shall consider in later chapters. In the literature, the notations $(\mathbf{x}, \mathbf{y})$ and $\mathbf{x}'\mathbf{y}$ are equally common, so it may be well to bear them both in mind.

Proposition 7.2

For every $\mathbf{x} \in \mathbf{R}^n$, $\mathbf{y} \in \mathbf{R}^n$, $\mathbf{z} \in \mathbf{R}^n$, *and* $\lambda \in \mathbf{R}$, *the following assertions hold:*

   (i) $(\mathbf{x}, \mathbf{y}) = (\mathbf{y}, \mathbf{x})$;

   (ii) $(\mathbf{x}, \mathbf{y} + \mathbf{z}) = (\mathbf{x}, \mathbf{y}) + (\mathbf{x}, \mathbf{z})$;

   (iii) $(\mathbf{x}, \lambda\mathbf{y}) = \lambda(\mathbf{x}, \mathbf{y})$;

   (iv) $(\mathbf{x}, \mathbf{x}) \geq 0$;

   (v) $(\mathbf{x}, \mathbf{x}) = 0$ *iff* $\mathbf{x} = \mathbf{0}$.

PROOF:
Left to the reader. ∎

Two members, $\mathbf{x}$ and $\mathbf{y}$, of $\mathbf{R}^n$ are said to be **orthogonal** (to each other) iff

$$(\mathbf{x}, \mathbf{y}) = 0.$$

Note that the null element $\mathbf{0}$ is (trivially) orthogonal to every member of $\mathbf{R}^n$. It is also easy to see that the unit vectors $\mathbf{e}_1, \mathbf{e}_2, \ldots, \mathbf{e}_n$ are orthogonal to one another.

A brief digression: The word "orthogonal" suggests that a right angle is somehow involved. To illustrate this connection, let us consider the special case of the space $\mathbf{R}^2$. Let $\mathbf{x}$ and $\mathbf{y}$ be nonnull elements of $\mathbf{R}^2$, and assume that $\mathbf{x}$ and $\mathbf{y}$ are orthogonal. It is well known that elements of $\mathbf{R}^2$ can be described as points in the plane, using a fixed horizontal axis and a fixed vertical axis. If we do this with $\mathbf{x}$ and $\mathbf{y}$, and if we draw the segments connecting the points $\mathbf{x}$ and $\mathbf{y}$ with the origin, we shall find that these segments form a right angle. The reader may wish to carry out this task for some specific choices of the vectors $\mathbf{x}$ and $\mathbf{y}$. The situation in the space $\mathbf{R}^n$ is similar, although it is more difficult to visualize. Let $\mathbf{u}$ and $\mathbf{v}$ be nonnull elements of $\mathbf{R}^n$, and assume that $\mathbf{u}$ and $\mathbf{v}$ are orthogonal. Let M be the subspace spanned by $\{\mathbf{u}, \mathbf{v}\}$. As we shall see in Proposition 7.3, the space M is two-dimensional, and therefore one may look upon elements of M as points in the plane. (Indeed, M is often referred to as "the plane passing through $\mathbf{u}$, $\mathbf{v}$, and $\mathbf{0}$.") In this plane, the segment connecting $\mathbf{u}$ with $\mathbf{0}$ and the segment connecting $\mathbf{v}$ with $\mathbf{0}$ form a right angle.

Proposition 7.3

Let $\mathbf{x}_1, \mathbf{x}_2, \ldots, \mathbf{x}_k$ be nonnull elements of $\mathbf{R}^n$. If $\mathbf{x}_1, \mathbf{x}_2, \ldots, \mathbf{x}_k$ are pairwise orthogonal, i.e., if $(\mathbf{x}_i, \mathbf{x}_j) = 0$ whenever $i \neq j$, then the set $\{\mathbf{x}_1, \mathbf{x}_2, \ldots, \mathbf{x}_k\}$ is linearly independent.

PROOF:

Assume, contrary to the assertion, that the set $\{\mathbf{x}_1, \mathbf{x}_2, \ldots, \mathbf{x}_k\}$ is linearly dependent. Then, there exist real numbers $\alpha_1, \alpha_2, \ldots, \alpha_k$, not all zero, such that $\sum \alpha_i \mathbf{x}_i = \mathbf{0}$. Suppose that $\alpha_j \neq 0$, and take the inner product of $\sum \alpha_i \mathbf{x}_i$ with $\mathbf{x}_j$:

$$\left(\mathbf{x}_j, \sum_{i=1}^{n} \alpha_i \mathbf{x}_i\right) = (\mathbf{x}_j, \mathbf{0}) = 0.$$

It follows from parts (ii) and (iii) of Proposition 7.2 that

$$\sum_{i=1}^{n} \alpha_i (\mathbf{x}_j, \mathbf{x}_i) = 0.$$

But $(\mathbf{x}_j, \mathbf{x}_i) = 0$ for all $i \neq j$. Hence, it must be true that $\alpha_j(\mathbf{x}_j, \mathbf{x}_j) = 0$, and since $\alpha_j \neq 0$, we get $(\mathbf{x}_j, \mathbf{x}_j) = 0$. This contradicts part (v) of Proposition 7.2, for $\mathbf{x}_j$ is, by assumption, a nonnull element of $\mathbf{R}^n$. ∎

Corollary

Let $\mathbf{x}_1, \mathbf{x}_2, \ldots, \mathbf{x}_n$ be nonnull elements of $\mathbf{R}^n$, and assume that these elements are pairwise orthogonal. Then, $\langle \mathbf{x}_1, \mathbf{x}_2, \ldots, \mathbf{x}_n \rangle$ is a basis for $\mathbf{R}^n$.

PROOF:

This follows immediately from the foregoing proposition, in conjunction with Proposition 6.5. ∎

A basis for $\mathbf{R}^n$ whose components are pairwise orthogonal is called, naturally, an **orthogonal basis**. If $\langle \mathbf{x}_1, \mathbf{x}_2, \ldots, \mathbf{x}_n \rangle$ is an orthogonal basis for $\mathbf{R}^n$ and if, furthermore, $(\mathbf{x}_i, \mathbf{x}_i) = 1$ for every i, then we refer to $\langle \mathbf{x}_1, \mathbf{x}_2, \ldots, \mathbf{x}_n \rangle$ as an **orthonormal basis** for $\mathbf{R}^n$. The most common example of an orthonormal basis for $\mathbf{R}^n$ is the unit basis, **I**.

When working with orthonormal bases, it is convenient to make use of a notational convention known as the **Kronecker delta**, δ_{ij}. Given any two positive integers, i and j, we define

$$\delta_{ij} = 1 \quad \text{if} \quad i = j$$
$$= 0 \quad \text{if} \quad i \neq j.$$

Thus, $\langle \mathbf{x}_1, \mathbf{x}_2, \ldots, \mathbf{x}_n \rangle$ is an orthonormal basis for $\mathbf{R}^n$ iff

$$(\mathbf{x}_i, \mathbf{x}_j) = \delta_{ij} \quad \text{for} \quad i = 1, 2, \ldots, n \quad \text{and} \quad j = 1, 2, \ldots, n.$$

Orthonormal bases have many pleasant properties, and it would therefore be useful to be able to find an orthonormal basis for every linear subspace of $\mathbf{R}^n$. The following proposition says that this can always be done.

Proposition 7.4

Let M be an m-dimensional subspace of $\mathbf{R}^n$. There exists an orthonormal basis for M.

PROOF:

Let $\langle \mathbf{x}_1, \mathbf{x}_2, \ldots, \mathbf{x}_m \rangle$ be a basis for M. We shall now find another m-tuple of members of M, $\langle \mathbf{y}_1, \mathbf{y}_2, \ldots, \mathbf{y}_m \rangle$, with the following properties: (i) For each positive integer $k \leq m$, $\mathbf{y}_k$ is a linear combination of $\mathbf{x}_1$, $\mathbf{x}_2, \ldots, \mathbf{x}_k$. (ii) $(\mathbf{y}_i, \mathbf{y}_j) = \delta_{ij}$ for all i and j. If we succeed in doing this, then the m-tuple $\langle \mathbf{y}_1, \mathbf{y}_2, \ldots, \mathbf{y}_m \rangle$ will surely be an orthonormal basis for M.

The construction proceeds as follows: Consider a positive integer k, $k < m$, and assume that we have already found a k-tuple $\langle \mathbf{y}_1, \mathbf{y}_2, \ldots, \mathbf{y}_k \rangle$ of linear combinations of $\mathbf{x}_1, \mathbf{x}_2, \ldots, \mathbf{x}_k$, such that $(\mathbf{y}_i, \mathbf{y}_j) = \delta_{ij}$. We shall now use this k-tuple to produce a new vector, $\mathbf{y}_{k+1}$, such that the $(k+1)$-tuple $\langle \mathbf{y}_1, \mathbf{y}_2, \ldots, \mathbf{y}_{k+1} \rangle$ will have the desired properties. To this end, define $\mathbf{z}_1$ by

$$\mathbf{z}_1 = \mathbf{x}_{k+1} - (\mathbf{x}_{k+1}, \mathbf{y}_1)\mathbf{y}_1.$$

It is easy to verify, by direct computation, that $(z_1, y_1) = 0$. Further-more, it is claimed that $z_1 \neq 0$. To see why, assume $z_1 = 0$. Then, by the equation defining z_1,

$$x_{k+1} = \alpha y_1$$

for some real number α. But this is impossible, since y_1 is a linear com-bination of $x_1, x_2, \ldots, x_k$, so that the foregoing equation makes x_{k+1} a linear combination of $x_1, x_2, \ldots, x_k$, contradicting the linear inde-pendence of the x's. Now define z_2 by

$$z_2 = z_1 - (z_1, y_2)y_2.$$

Again by direct computation, we find that $(z_2, y_1) = 0$ as well as $(z_2, y_2) = 0$. Also, $z_2 \neq 0$. For if $z_2 = 0$, then we have

$$z_1 = (z_1, y_2)y_2,$$

and, substituting for z_1, we get

$$x_{k+1} - \alpha_1 y_1 = \alpha_2 y_2$$

for some real numbers α_1 and α_2, and this, once again, contradicts the linear independence of the x's. Now, if we define z_3 by

$$z_3 = z_2 - (z_2, y_3)y_3,$$

we find that z_3 is nonnull and that it is orthogonal to each of the vectors y_1, y_2, and y_3. We may continue in this manner until we finally arrive at a vector z_k that is nonnull and is orthogonal to each of the vectors $y_1, y_2, \ldots, y_k$. It remains only to define y_{k+1} by writing

$$y_{k+1} = (1/\sqrt{(z_k, z_k)})z_k,$$

and the construction is complete. The proof, however, is not yet quite complete. All we have shown in the foregoing argument was that if k steps can be taken, then $k + 1$ steps can also be taken. We still have to argue that the first step can, in fact, be taken. But this is immediate, since the first step consists merely of setting $y_1 = (1/\sqrt{(x_1, x_1)})x_1$. ∎

The method used in the foregoing proof to convert a basis into an ortho-normal basis is known as **Gram-Schmidt orthogonalization**.

Consider an arbitrary nonempty subset S of R^n. The set of all members of $\mathbf{R}^n$ that are orthogonal to every member of S is called the **orthogonal complement** of S, and it is denoted $S^\perp$:

$$S^\perp = \{\mathbf{x} \mid (\mathbf{x}, \mathbf{y}) = 0 \quad \text{for every} \quad \mathbf{y} \in S\}.$$

Proposition 7.5

Let S be a nonempty subset of $\mathbf{R}^n$. Then, $S^\perp$ is a linear subspace of $\mathbf{R}^n$.

PROOF:

Left to the reader. ∎

In what sense is $S^\perp$ a "complement" of S? The following proposition provides a possible answer to this question. It says that if M is an m-dimensional subspace of $\mathbf{R}^n$, then $M^\perp$ has the complementary dimension, namely $n - m$.

Proposition 7.6

Let M be an m-dimensional subspace of $\mathbf{R}^n$. Then $\dim M^\perp = n - m$.

PROOF:

Let $\mathbf{A} = \langle \mathbf{x}_1, \mathbf{x}_2, \ldots, \mathbf{x}_m \rangle$ and $\mathbf{B} = \langle \mathbf{y}_1, \mathbf{y}_2, \ldots, \mathbf{y}_k \rangle$ be orthonormal bases for M and $M^\perp$, respectively. Such orthonormal bases exist by Proposition 7.4. Now define $p = m + k$. We wish to show that $p = n$. Consider the p-tuple $\langle \mathbf{x}_1, \mathbf{x}_2, \ldots, \mathbf{x}_m, \mathbf{y}_1, \mathbf{y}_2, \ldots, \mathbf{y}_k \rangle$, and call it $\mathbf{C}$ for short. The components of $\mathbf{C}$ are surely pairwise orthogonal and nonnull, and therefore they must be linearly independent. Hence, p cannot exceed n. Suppose, on the other hand, that $p < n$. Then, there exists an element $\mathbf{u}$ of $\mathbf{R}^n$ that is not a linear combination of the components of $\mathbf{C}$. Using Gram-Schmidt orthogonalization, we can convert $\mathbf{u}$ into a nonnull element, say $\mathbf{v}$, which is orthogonal to every component of $\mathbf{C}$. In particular, $\mathbf{v}$ is orthogonal to $\mathbf{x}_1, \mathbf{x}_2, \ldots, \mathbf{x}_m$, which implies that it is orthogonal to every member of M. In other words, $\mathbf{v}$ must belong to $M^\perp$. Now $\mathbf{v}$ is orthogonal also to $\mathbf{y}_1, \mathbf{y}_2, \ldots, \mathbf{y}_k$, which implies that it is orthogonal to every member of $M^\perp$. This means, since $\mathbf{v}$ itself belongs to $M^\perp$, that $\mathbf{v}$ is orthogonal to itself, i.e., that $(\mathbf{v}, \mathbf{v}) = 0$. But, by Proposition 7.2, we now have $\mathbf{v} = \mathbf{0}$, which is a contradiction. ∎

Two further properties of orthogonal complements, which will be of use later on, are described in the following propositions.

Proposition 7.7

Let M be a subspace of $\mathbf{R}^n$, and let $\mathbf{z}$ be an arbitrary member of $\mathbf{R}^n$. Then, $\mathbf{z}$ can be written as a sum, $\mathbf{z} = \mathbf{x} + \mathbf{y}$, where $\mathbf{x} \in M$ and $\mathbf{y} \in M^\perp$.

PROOF:

Let $\mathbf{B} = \langle x_1, \ldots, x_n \rangle$ be an orthonormal basis for $\mathbf{R}^n$, whose first m components constitute a basis for M, and whose last $n - m$ components constitute a basis for $M^\perp$. (In the course of proving Proposition 7.6, we have seen that such a basis certainly exists.) Expressing $\mathbf{z}$ in terms of this basis leads to the desired result. ∎

Proposition 7.8

Let M be a subspace of $\mathbf{R}^n$. Then, $(M^\perp)^\perp = M$.

PROOF:

Let $\mathbf{z}$ be a member of M. Then, for every member $\mathbf{y}$ of $M^\perp$, it is true, by definition, that $(\mathbf{z}, \mathbf{y}) = 0$. In other words, $\mathbf{z}$ is orthogonal to every member of $M^\perp$, and therefore $\mathbf{z}$ belongs to $(M^\perp)^\perp$. Conversely, suppose $\mathbf{z} \notin M$. Since M is a subspace, we may assume $\mathbf{z} \neq \mathbf{0}$. By Proposition 7.7, we have $\mathbf{z} = \mathbf{x} + \mathbf{y}$, where $\mathbf{x} \in M$, $\mathbf{y} \in M^\perp$, and $\mathbf{y} \neq \mathbf{0}$. Taking the inner product of $\mathbf{z}$ with $\mathbf{y}$, we get

$$(\mathbf{z}, \mathbf{y}) = (\mathbf{x}, \mathbf{y}) + (\mathbf{y}, \mathbf{y}) = (\mathbf{y}, \mathbf{y}) \neq 0.$$

In other words, $\mathbf{y}$ is a member of $M^\perp$ to which $\mathbf{z}$ is not orthogonal. Hence, $\mathbf{z} \notin (M^\perp)^\perp$. ∎

PROBLEMS

7.1. Let M be a subspace of $\mathbf{R}^n$. Show that there exists a linear transformation on $\mathbf{R}^n$ whose null space is M.

7.2. Prove Propositions 7.2 and 7.5.

7.3. Show that T is a linear transformation on $\mathbf{R}^n$ to $\mathbf{R}$ iff there exists an element $\mathbf{t}$ of $\mathbf{R}^n$ such that

$$T(\mathbf{x}) = (\mathbf{t}, \mathbf{x}) \quad \text{for every} \quad \mathbf{x} \in \mathbf{R}^n.$$

7.4. Let $\mathbf{x}$ and $\mathbf{y}$ be elements of $\mathbf{R}^n$. An important inequality connecting the inner products $(\mathbf{x}, \mathbf{y})$, $(\mathbf{x}, \mathbf{x})$, and $(\mathbf{y}, \mathbf{y})$, known as **Schwarz's inequality**, says that

$$(\mathbf{x}, \mathbf{y})^2 \leqq (\mathbf{x}, \mathbf{x})(\mathbf{y}, \mathbf{y}).$$

Prove this inequality. [Hint: Start by evaluating the inner product $(\mathbf{x} - \lambda\mathbf{y}, \mathbf{x} - \lambda\mathbf{y})$, where λ is an arbitrary real number.]

7.5. Let $\langle \mathbf{x}_1, \mathbf{x}_2, \ldots, \mathbf{x}_n \rangle$ be an orthonormal basis for $\mathbf{R}^n$. Show that for every $\mathbf{y} \in \mathbf{R}^n$ and $\mathbf{z} \in \mathbf{R}^n$,

$$(\mathbf{y}, \mathbf{z}) = \sum_{i=1}^{n} (\mathbf{y}, \mathbf{x}_i)(\mathbf{z}, \mathbf{x}_i).$$

7.6. Consider a subspace M of $\mathbf{R}^n$ and an element $\mathbf{y}$ of $\mathbf{R}^n$. Assume that $\mathbf{y} \notin M$. Let $\mathbf{z}$ be an element of M such that

$$(\mathbf{y} - \mathbf{z}) \in M^{\perp}.$$

Show that the following inequality is valid:

$$(\mathbf{y} - \mathbf{z}, \mathbf{y} - \mathbf{z}) \leqq (\mathbf{y} - \mathbf{x}, \mathbf{y} - \mathbf{x}) \quad \text{for every} \quad \mathbf{x} \in M.$$

What is the geometrical interpretation of this inequality, say, in the special case $n = 2$?

8 Linear Transformations
and Matrices

This chapter is devoted to linear transformations on finite-dimensional spaces. Since a finite-dimensional space is isomorphic to a space of the $\mathbf{R}^n$ type, we are entitled to carry out our investigation in terms of linear transformations on $\mathbf{R}^n$ to $\mathbf{R}^m$. Our first proposition is a kind of "law of preservation of dimension." It says, roughly speaking, that the dimension of the image of an n-dimensional space under a linear transformation cannot exceed n, and if it is less than n, then the "loss" of dimension is captured in the null space of the transformation.

Proposition 8.1

Let T be a linear transformation on $\mathbf{R}^n$ to $\mathbf{R}^m$, and let $\mathscr{R}(T)$ and $\mathscr{N}(T)$ be, respectively, the range and the null space of T. Then,

$$dim \, \mathscr{R}(T) + dim \, \mathscr{N}(T) = n.$$

PROOF:

Assume first that T has a trivial null space, so that $dim \, \mathscr{N}(T) = 0$. Then, by Proposition 5.4, T is an isomorphism on $\mathbf{R}^n$ onto $\mathscr{R}(T)$. Thus, $\mathscr{R}(T)$ must be n-dimensional, as asserted. Now suppose that $dim \, \mathscr{N}(T) > 0$, and let $\langle \mathbf{x}_1, \mathbf{x}_2, \ldots, \mathbf{x}_k \rangle$ be a basis for $\mathcal{N}(T)$. By Proposition 6.5, there exists a basis $\langle \mathbf{x}_1, \mathbf{x}_2, \ldots, \mathbf{x}_n \rangle$ for $\mathbf{R}^n$, whose first k components coincide with our basis for $\mathcal{N}(T)$. We shall establish the desired result by showing that $\langle T(\mathbf{x}_{k+1}), \ldots, T(\mathbf{x}_n) \rangle$ is a basis for $\mathcal{R}(T)$. First, consider an arbitrary member $\mathbf{y}$ of $\mathcal{R}(T)$. By definition, there exists an $\mathbf{x} \in \mathbf{R}^n$ such that $\mathbf{y} = T(\mathbf{x})$. Let $\langle \alpha_1, \alpha_2, \ldots, \alpha_n \rangle$ be the n-tuple of coordinates of $\mathbf{x}$ relative to the basis $\langle \mathbf{x}_1, \mathbf{x}_2, \ldots, \mathbf{x}_n \rangle$:

$$\mathbf{x} = \sum_{i=1}^{n} \alpha_i \mathbf{x}_i.$$

Now, by the properties of a linear transformation, we have

$$\mathbf{y} = T(\mathbf{x}) = \sum_{i=1}^{n} \alpha_i\, T(\mathbf{x}_i).$$

But $T(\mathbf{x}_i) = \mathbf{0}$ for $i = 1, 2, \ldots, k$, and therefore,

$$\mathbf{y} = \sum_{i=k+1}^{n} \alpha_i T(\mathbf{x}_i).$$

Hence, $\langle T(\mathbf{x}_{k+1}), \ldots, T(\mathbf{x}_n) \rangle$ spans $\mathcal{R}(T)$. To establish linear independence, let us assume the contrary, namely that there exist $n - k$ real numbers $\beta_{k+1}, \beta_{k+2}, \ldots, \beta_n$, not all zero, such that

$$\sum_{i=k+1}^{n} \beta_i T(\mathbf{x}_i) = \mathbf{0}.$$

It follows from the properties of a linear transformation that

$$T\left(\sum_{i=k+1}^{n} \beta_i \mathbf{x}_i\right) = \mathbf{0},$$

so that $\sum \beta_i \mathbf{x}_i$ must be a member of $\mathscr{N}(T)$, and therefore it must be a linear combination of $\mathbf{x}_1, \mathbf{x}_2, \ldots, \mathbf{x}_k$:

$$\sum_{i=k+1}^{n} \beta_i \mathbf{x}_i = \sum_{i=1}^{k} \gamma_i \mathbf{x}_i$$

for some real numbers $\gamma_1, \gamma_2, \ldots, \gamma_k$. This last equation contradicts the linear independence of $\mathbf{x}_1, \mathbf{x}_2, \ldots, \mathbf{x}_n$. ∎

The dimension of the range of a linear transformation is referred to as its **rank**, and the dimension of its null space is referred to as its **nullity**. A linear transformation whose nullity is zero (that is, a linear transformation whose null space is trivial) is referred to as being **of full rank**. The rank of a linear transformation T is denoted *rank(T)*.

Proposition 8.2

Let T_1 be a linear transformation on $\mathbf{R}^n$ to $\mathbf{R}^m$, and let T_2 be a linear transformation on $\mathbf{R}^m$ to $\mathbf{R}^s$. Define T to be the composition of T_1 and T_2. (By Proposition 5.6, T is a linear transformation on $\mathbf{R}^n$ to $\mathbf{R}^s$.) Then the following inequalities are valid:

$$rank\ (T) \leqq rank\ (T_1)$$
$$rank\ (T) \leqq rank\ (T_2).$$

PROOF:
From the definition of T, it follows immediately that

$$\mathscr{N}(T) \supset \mathscr{N}(T_1)$$

and that

$$\mathscr{R}(T) \subset \mathscr{R}(T_2).$$

The first of these two assertions implies that the nullity of T is greater than or equal to the nullity of T_1, so that, using Proposition 8.1, we conclude that $rank\ (T) \leqq rank\ (T_1)$. From the second assertion it follows directly that $rank\ (T) \leqq rank\ (T_2)$. ∎

Our next task will be to find an efficient way of characterizing a linear transformation on $\mathbf{R}^n$ to $\mathbf{R}^m$. It will be convenient, carrying out this task, to give a special name to n-tuples whose components are members of $\mathbf{R}^m$. An n-tuple of the form $\langle \mathbf{a}_1, \mathbf{a}_2, \ldots, \mathbf{a}_n \rangle$, where $\mathbf{a}_1, \mathbf{a}_2, \ldots, \mathbf{a}_n$ are members of $\mathbf{R}^m$ is called a **real** $m \times n$ **matrix** (read: real m by n matrix). We shall use bold-faced Roman characters to denote real matrices.

It is often convenient to write a real $m \times n$ matrix as a rectangular array of real numbers. For example, the matrix $\mathbf{A} = \langle \mathbf{a}_1, \ldots, \mathbf{a}_n \rangle$ may be written as follows:

$$\mathbf{A} = \begin{bmatrix} a_{11} & a_{12} & a_{13} & \cdots & a_{1n} \\ a_{21} & a_{22} & a_{23} & \cdots & a_{2n} \\ \cdot & \cdot & \cdot & & \\ \cdot & \cdot & \cdot & & \\ \cdot & \cdot & \cdot & & \\ a_{m1} & a_{m2} & a_{m3} & \cdots & a_{mn} \end{bmatrix}$$

where the real number a_{ij} is the i-th component of the vector $\mathbf{a}_j$. a_{ij} is often referred to as the ij-th **entry** of $\mathbf{A}$. The vector $\mathbf{a}_j$, given by

$$\mathbf{a}_j = \begin{bmatrix} a_{1j} \\ a_{2j} \\ \cdot \\ \cdot \\ \cdot \\ a_{mj} \end{bmatrix},$$

is referred to as the j-th **column** of $\mathbf{A}$. Often, one is also interested in the n-tuple

$$\langle a_{i1}, a_{i2}, \ldots, a_{in} \rangle,$$

which is referred to as the i-th **row** of $\mathbf{A}$. If $\mathbf{A}$ is a real $m \times n$ matrix with entries a_{ij}, then it is common to write

$$\mathbf{A} = [a_{ij}]_{\substack{i=1,\ldots,m \\ j=1,\ldots,n}}$$

and when there is no danger of confusion, one writes simply

$$\mathbf{A} = [a_{ij}].$$

The adjective "real" in the phrase "real $m \times n$ matrix" refers to the fact that the entries of the matrix are real numbers. In general, a matrix is a rectangular array whose entries are members of a field. In the present volume, however, we shall restrict our attention to the case where this field is the real number system, $\mathbf{R}$.

Let $\mathbf{A}$ be a real $m \times n$ matrix. It is important to remember that in the case $n = 1$, the matrix $\mathbf{A}$ is simply an element of the space $\mathbf{R}^m$. Similarly, every element of $\mathbf{R}^m$ may be looked upon as a real $m \times 1$ matrix. Thus, all that will be said here about real $m \times n$ matrices applies, as a special case, to members of $\mathbf{R}^m$.

Consider two real $m \times n$ matrices, $\mathbf{A} = [a_{ij}]$ and $\mathbf{B} = [b_{ij}]$. There is a natural way to define the sum of $\mathbf{A}$ and $\mathbf{B}$, namely,

$$\mathbf{A} + \mathbf{B} \underset{\text{def}}{=} [a_{ij} + b_{ij}],$$

and thus have *addition* defined in the set of all real $m \times n$ matrices. Similarly, if $\mathbf{A} = [a_{ij}]$ is a real $m \times n$ matrix, and if λ is a real number, then one can define

$$\lambda \mathbf{A} \underset{\text{def}}{=} [\lambda a_{ij}]$$

and thereby introduce multiplication by a real number into the set of all real $m \times n$ matrices. As might be expected, the set of all real $m \times n$ matrices, together with addition and multiplication by a real number as defined here, constitutes a real linear space.

We now proceed to define a further operation on matrices, to be called **matrix multiplication**. Let $\mathbf{A} = [a_{ij}]$ be a real $m \times n$ matrix, and let $\mathbf{B} = [b_{ij}]$ be a real $n \times s$ matrix. Note that $\mathbf{B}$ has as many rows as $\mathbf{A}$ has columns. We define the product, $\mathbf{AB}$, by

$$\mathbf{AB} \underset{\text{def}}{=} \left[\sum_{k=1}^{n} a_{ik} b_{kj} \right]_{\substack{i=1,\ldots,m \\ j=1,\ldots,s.}}$$

In words, $\mathbf{AB}$ is the real $m \times s$ matrix whose ij-th entry is the inner product of the i-th row of $\mathbf{A}$ with the j-th column of $\mathbf{B}$. At first blush, this definition looks somewhat contrived. Its significance will emerge, however, in Propositions 8.3 and 8.4 below. It may be well to emphasize again that the product $\mathbf{AB}$ is well defined only if $\mathbf{B}$ has exactly as many rows as $\mathbf{A}$ has columns. From the fact that the product $\mathbf{AB}$ is well defined it certainly does not follow that the product $\mathbf{BA}$ is well defined. Moreover, even if both products, $\mathbf{AB}$ and $\mathbf{BA}$, are well defined, they need not be equal. Thus, in speaking about the product of the matrix $\mathbf{A}$ and the matrix $\mathbf{B}$, we must be careful to specify the order in which the two matrices appear in the product. To this end, let us introduce the following terminology: In the product $\mathbf{AB}$, the matrix $\mathbf{A}$ is said to **premultiply** the matrix **B**, and the matrix **B** is said to **postmultiply** the matrix **A**. In other words, "premultiplication" is short for "multiplication on the left" and "postmultiplication" is short for "multiplication on the right."

We are now ready to resume the discussion of linear transformations on the space $\mathbf{R}^n$ to the space $\mathbf{R}^m$.

Consider an arbitrary function f on $\mathbf{R}^n$ to $\mathbf{R}^m$. In general, in order to specify the function f completely, one has to use the definition of f and write down the set of all pairs of the form $\langle \mathbf{x}, \mathbf{y} \rangle$, where $\mathbf{x}$ runs throughout $\mathbf{R}^n$ and $\mathbf{y}$ is the image of $\mathbf{x}$ under f. In other words, in order to characterize f, one must, in general, specify the action of f on each member of $\mathbf{R}^n$. In the case of a linear transformation, however, things are much simpler. The properties of a linear transformation make it possible to characterize it completely by describing its action only on certain selected members of $\mathbf{R}^n$. Specifically, if T is a linear transformation on $\mathbf{R}^n$ to $\mathbf{R}^m$, and if $\mathbf{B} = \langle \mathbf{x}_1, \mathbf{x}_2, \ldots, \mathbf{x}_n \rangle$ is a basis for $\mathbf{R}^n$, then it is clear that the action of T on $\mathbf{x}_1, \mathbf{x}_2, \ldots, \mathbf{x}_n$ determines its action on all members of $\mathbf{R}^n$. In fact, if $\mathbf{y}_1, \mathbf{y}_2, \ldots, \mathbf{y}_n$ are the respective images of $\mathbf{x}_1$, $\mathbf{x}_2, \ldots, \mathbf{x}_n$ under T, and if $\mathbf{x}$ is an arbitrary member of $\mathbf{R}^n$, then the image of $\mathbf{x}$ under T is given by $\alpha_i \mathbf{y}_i$, where $\langle \alpha_1, \alpha_2, \ldots, \alpha_n \rangle$ is the n-tuple of coordinates of $\mathbf{x}$ relative to the basis $\mathbf{B}$. Now, $\mathbf{B}$ here stands for an arbitrary basis for $\mathbf{R}^n$, and as long as we have a choice as to which basis to select, we might as well pick the most convenient one, namely the unit basis, $\mathbf{I}$. This leads to the following definition:

Let T be a linear transformation on $\mathbf{R}^n$ to $\mathbf{R}^m$, and let $\mathbf{a}_1, \mathbf{a}_2, \ldots, \mathbf{a}_n$ be the respective images of the unit vectors $\mathbf{e}_1, \mathbf{e}_2, \ldots, \mathbf{e}_n$ under T. Then, the real $m \times n$ matrix $\langle \mathbf{a}_1, \mathbf{a}_2, \ldots, \mathbf{a}_n \rangle$ is referred to as the **natural matrix** of T.

The word "natural" was used in the foregoing definition because we have chosen to work with the *unit basis* (also called the *natural basis*) for $\mathbf{R}^n$. Furthermore, our definition conceals an implicit choice of a particular basis for the space $\mathbf{R}^m$ as well, this basis being, once again, the unit basis. In general, if T is a linear transformation on $\mathbf{R}^n$ to $\mathbf{R}^m$, then there is a real $m \times n$ matrix associated with T for each choice of bases for $\mathbf{R}^n$ and $\mathbf{R}^m$. Specifically, let $\mathbf{B} = \langle \mathbf{x}_1, \mathbf{x}_2, \ldots, \mathbf{x}_n \rangle$ and $\mathbf{C} = \langle \mathbf{y}_1, \mathbf{y}_2, \ldots, \mathbf{y}_m \rangle$ be, respectively, a basis for $\mathbf{R}^n$ and a basis for $\mathbf{R}^m$, and let T be a linear transformation on $\mathbf{R}^n$ to $\mathbf{R}^m$. Now define $\mathbf{z}_1, \mathbf{z}_2, \ldots, \mathbf{z}_n$ to be the respective images of $\mathbf{x}_1, \mathbf{x}_2, \ldots, \mathbf{x}_n$ under T, and let $\mathbf{v}_1, \mathbf{v}_2, \ldots, \mathbf{v}_n$ be the respective m-tuples of coordinates of $\mathbf{z}_1, \mathbf{z}_2$, $\ldots, \mathbf{z}_n$ relative to the basis $\mathbf{C}$. Then, the real matrix $\langle \mathbf{v}_1, \mathbf{v}_2, \ldots, \mathbf{v}_n \rangle$ is referred to as **the matrix of T relative to the bases B and C**. As has already been mentioned, in the case where $\mathbf{B}$ and $\mathbf{C}$ are, respectively, the unit basis for $\mathbf{R}^n$ and the unit basis for $\mathbf{R}^m$, the matrix of T relative to $\mathbf{B}$ and $\mathbf{C}$ will be referred to, for short, as the *natural* matrix of T. In the sequel, we shall describe linear transformations mostly in terms of their natural matrices, rather than in terms of their matrices relative to some arbitrary bases. It should be remarked, however, that most of the assertions concerning the natural matrix of a linear transformation have analogues in terms of the matrix of a linear transformation relative to a given pair of arbitrary bases.

Proposition 8.3

Let T be a linear transformation on $\mathbf{R}^n$ to $\mathbf{R}^m$, and let $\mathbf{A}$ be a real $m \times n$ matrix. $\mathbf{A}$ is the natural matrix of T iff, for each $\mathbf{x} \in \mathbf{R}^n$,

$$T(\mathbf{x}) = \mathbf{Ax},$$

where $\mathbf{Ax}$ is the product obtained when the real $n \times 1$ matrix $\mathbf{x}$ is premultiplied by the real $m \times n$ matrix $\mathbf{A}$.

PROOF:

Let $\mathbf{a}_j$ be the j-th column of $\mathbf{A}$, and let $\mathbf{x}$ be an arbitrary member of $\mathbf{R}^n$, with components $x_1, x_2, \ldots, x_n$. It follows immediately from the definition of the product of two matrices that

$$\mathbf{Ax} = \sum_{j=1}^{n} x_j \mathbf{a}_j.$$

Now, if $\mathbf{A}$ is the natural matrix of T, then we have

$$T(\mathbf{x}) = T\left(\sum_{j=1}^{n} x_j \mathbf{e}_j\right) = \sum_{j=1}^{n} x_j T(\mathbf{e}_j) = \sum_{j=1}^{n} x_j \mathbf{a}_j,$$

by the properties of a linear transformation. Conversely, assume that $T(\mathbf{x}) = \mathbf{Ax}$ for every $\mathbf{x} \in \mathbf{R}^n$. Then, in particular, $T(\mathbf{e}_i) = \mathbf{Ae}_i$ for $i = 1, 2, \ldots, n$. Now, the product $\mathbf{Ae}_i$ is readily seen to be equal to $\mathbf{a}_i$, the i-th column of $\mathbf{A}$. Thus, $\mathbf{A}$ is the natural matrix of T. ∎

Using this proposition, we are able to translate statements about linear transformations into statements about their natural matrices. Since all the information about a linear transformation is contained in its natural matrix, it will be advantageous to make full use of this property and, in effect, to conduct the study of linear transformations in terms of their natural matrices. Some of the terminology that was introduced earlier for linear transformations will also be used for analogous concepts in terms of matrices. Specifically, let $\mathbf{A}$ be a real $m \times n$ matrix and consider the set

$$\{\mathbf{y} \mid \mathbf{y} \in \mathbf{R}^m, \mathbf{y} = \mathbf{Ax} \quad \textit{for some} \quad \mathbf{x} \in \mathbf{R}^n\}.$$

Rather than refer to this set as the range of the linear transformation whose natural matrix is $\mathbf{A}$, we shall refer to it simply as the range of $\mathbf{A}$, and we shall denote it $\mathscr{R}(\mathbf{A})$. Similarly, rather than refer to the dimension of $\mathscr{R}(\mathbf{A})$ as the rank of the linear transformation whose natural matrix is $\mathbf{A}$, we shall refer to it as the rank of $\mathbf{A}$, to be denoted *rank* $\mathbf{A}$. Finally, the set

$$\{\mathbf{x} \,|\, \mathbf{x} \in \mathbf{R}^n, \, \mathbf{A}\mathbf{x} = \mathbf{0}\}$$

will be referred to as the null space of $\mathbf{A}$, denoted $\mathscr{N}(\mathbf{A})$.

The composition of two linear transformations is, by Proposition 5.5, again a linear transformation. Since all the information concerning a linear transformation is contained in its natural matrix, we expect to find a connection between the natural matrix of the composition and the natural matrices of the transformations that are being composed. This connection is a very simple one, and it is given in the following proposition:

Proposition 8.4

Consider a linear transformation T on $\mathbf{R}^n$ to $\mathbf{R}^m$ and a linear transformation Q on $\mathbf{R}^m$ to $\mathbf{R}^s$, and let $\mathbf{A}$ and $\mathbf{B}$ be the respective natural matrices of T and Q. Then, the natural matrix of the composition $Q \circ T$ is the product $\mathbf{B}\mathbf{A}$.

PROOF:

Let $\mathbf{x}$ be an arbitrary member of $\mathbf{R}^n$. Then, in view of Proposition 8.3, all that has to be shown is

$$\mathbf{B}(\mathbf{A}\mathbf{x}) = (\mathbf{B}\mathbf{A})\mathbf{x}.$$

More generally, our proposition will be proved if matrix multiplication can be shown to be associative. But matrix multiplication *is* associative, as may be verified by direct computation. ∎

Let T be an *invertible* linear transformation on $\mathbf{R}^n$ onto $\mathbf{R}^m$. Then, it must be true that $m = n$, so that the natural matrix of T, call it $\mathbf{A}$, is an $n \times n$ matrix. The natural matrix of T^{-1} is also an $n \times n$ matrix and, as might be expected, it is referred to as the inverse of $\mathbf{A}$, and it is denoted $\mathbf{A}^{-1}$. In view of Proposition 8.4 and the properties of invertible functions, the connection between $\mathbf{A}$ and $\mathbf{A}^{-1}$ is given by

$$\mathbf{A}\mathbf{A}^{-1} = \mathbf{A}^{-1}\mathbf{A} = \mathbf{I},$$

where $\mathbf{I}$ is the natural matrix of the *identity* on $\mathbf{R}^n$. Now, the images of the unit vectors under the identity are again the unit vectors, so $\mathbf{I}$ is the matrix whose i-th column is the i-th unit vector, for $i = 1, 2, \ldots, n$:

$$\mathbf{I} = \langle \mathbf{e}_1, \mathbf{e}_2, \ldots, \mathbf{e}_n \rangle.$$

This matrix is referred to as the $n \times n$ **identity matrix**. Note that the very same symbol, **I**, with the very same definition, appeared first in Chapter 7, where it was named the *unit basis*. It is hoped that the reader will not be disturbed by this. The unit basis and the identity matrix are, indeed, one and the same thing, although we have given it two different names in order to achieve two different emphases.

PROBLEMS

8.1. What is the dimension of the linear space whose elements are real $m \times n$ matrices? Prove your assertion.

8.2. Let T and Q be two linear transformations on $\mathbf{R}^n$ to $\mathbf{R}^m$, and let λ be a real number. Define $T + Q$ and λT by the formulas

$$\left. \begin{array}{c} (T + Q)(\mathbf{x}) = T(\mathbf{x}) + Q(\mathbf{x}) \\ (\lambda T)(\mathbf{x}) = \lambda T(\mathbf{x}) \end{array} \right\} \quad \text{for every} \quad \mathbf{x} \in \mathbf{R}^n.$$

With these definitions, the set of all linear transformations on $\mathbf{R}^n$ to $\mathbf{R}^m$ becomes a real linear space. Show that this space is *isomorphic* to the space of all real $m \times n$ matrices.

8.3. Let T be a linear transformation on $\mathbf{R}^3$ to $\mathbf{R}^3$. Is it possible for the range of T and for the null space of T to coincide? Prove your assertion.

8.4. Each of the following four equations defines a linear transformation on $\mathbf{R}^2$ to $\mathbf{R}^2$ (all four linear transformations are denoted T):

(i) $T \begin{bmatrix} \alpha \\ \beta \end{bmatrix} = \begin{bmatrix} \beta \\ \alpha \end{bmatrix}$

(ii) $T \begin{bmatrix} \alpha \\ \beta \end{bmatrix} = \begin{bmatrix} \alpha + \beta \\ \alpha - \beta \end{bmatrix}$

(iii) $T \begin{bmatrix} \alpha \\ \beta \end{bmatrix} = \begin{bmatrix} (\alpha + \beta)/2 \\ (\alpha + \beta)/2 \end{bmatrix}$

(iv) $T \begin{bmatrix} \alpha \\ \beta \end{bmatrix} = \begin{bmatrix} 0 \\ \alpha \end{bmatrix}$

where α and β are arbitrary real numbers. In each case, find the range, the null space, and the natural matrix of T. Exhibit the range and the null space of T in a diagram.

8.5. Consider the following matrices:

$$A = \begin{bmatrix} 3 & 2 \\ -1 & 0 \end{bmatrix};$$

$$B = \begin{bmatrix} 2 & 1 \\ 2 & 2 \end{bmatrix};$$

$$C = \begin{bmatrix} -3 & 1 \\ -6 & 2 \end{bmatrix};$$

$$D = \begin{bmatrix} 3 & -3 \\ 1 & 2 \end{bmatrix};$$

$$E = \begin{bmatrix} 5 & 1 \\ 0 & 0 \end{bmatrix}.$$

Find the ranges and the null spaces of the matrices **A, B, C, D, E, AB, ABC, ABCD,** and **ABCDE**. In each case, exhibit the range and the null space in a diagram.

9 The Transpose

In Chapter 8, it was stated that the set of all real $m \times n$ matrices, with addition and multiplication by a real number defined in the natural way, is a real linear space. One is therefore in a position to speak of linear transformations on the space of all real $m \times n$ matrices to the space of all real $r \times s$ matrices. Lest the reader be alarmed by this dreary prospect, let us hasten to add that we shall consider here only one such linear transformation.

Let $\mathbf{A}$ be a real $m \times n$ matrix:

$$\mathbf{A} = \begin{bmatrix} a_{11} & a_{12} & a_{13} & \ldots & a_{1n} \\ a_{21} & a_{22} & a_{23} & \ldots & a_{2n} \\ \cdot & \cdot & \cdot & & \cdot \\ \cdot & \cdot & \cdot & & \cdot \\ \cdot & \cdot & \cdot & & \cdot \\ a_{m1} & a_{m2} & a_{m3} & \ldots & a_{mn} \end{bmatrix}.$$

Then, the real $n \times m$ matrix

$$\begin{bmatrix} a_{11} & a_{21} & a_{31} & \ldots & a_{m1} \\ a_{12} & a_{22} & a_{32} & \ldots & a_{m2} \\ \cdot & \cdot & \cdot & & \cdot \\ \cdot & \cdot & \cdot & & \cdot \\ \cdot & \cdot & \cdot & & \cdot \\ a_{1n} & a_{2n} & a_{3n} & \ldots & a_{mn} \end{bmatrix},$$

i.e., the matrix whose columns are the rows of $\mathbf{A}$ and whose rows are the columns of $\mathbf{A}$, is called the **transpose** of $\mathbf{A}$, and it is denoted $\mathbf{A}'$. The function that associates $\mathbf{A}'$ with $\mathbf{A}$ for each real $m \times n$ matrix $\mathbf{A}$ is called **transposition**.

The following proposition lists four rather obvious properties of the transpose.

Proposition 9.1

Let $\mathbf{A}$ and $\mathbf{B}$ be real $m \times n$ matrices, and let $\mathbf{C}$ be a real $n \times s$ matrix. Furthermore, let λ be an arbitrary real number. Then,

(i) $(\mathbf{A} + \mathbf{B})' = \mathbf{A}' + \mathbf{B}'$;

(ii) $(\lambda\mathbf{A})' = \lambda\mathbf{A}'$;

(iii) $(\mathbf{A}')' = \mathbf{A}$;

(iv) $(\mathbf{AC})' = \mathbf{C}'\mathbf{A}'$.

PROOF:

The first three assertions are immediate, and the fourth may be verified by direct computation. ∎

Parts (i) and (ii) of this proposition entail that transposition is a linear transformation. That transposition is one-to-one and onto is obvious, so that it is, in fact, an *isomorphism* on the space of real $m \times n$ matrices onto the space of real $n \times m$ matrices. Part (iii) of the foregoing proposition tells us that the inverse of transposition is transposition, and part (iv) says that the transpose of the product of two matrices is the product of the transposes of these matrices in reverse order.

In Chapter 7, we noted that the inner product, $(\mathbf{x}, \mathbf{y})$, of two elements of $\mathbf{R}^n$, may also be written $\mathbf{x}'\mathbf{y}$. This notation arises from the fact that the inner product $(\mathbf{x}, \mathbf{y})$ is, indeed, the same thing as the product obtained by taking the real $n \times 1$ matrix $\mathbf{y}$ and premultiplying it by the transpose of the real $n \times 1$ matrix $\mathbf{x}$.

Consider a real $m \times n$ matrix $\mathbf{A}$. We know that $\mathbf{A}$ is the natural matrix of a linear transformation on $\mathbf{R}^n$ to $\mathbf{R}^m$ and that $\mathbf{A}'$ is the natural matrix of a linear transformation on $\mathbf{R}^m$ to $\mathbf{R}^n$. The following proposition explores the connection between these two linear transformations.

Proposition 9.2

Let $\mathbf{A}$ be a real $m \times n$ matrix. Then,

$$\mathscr{R}(\mathbf{A})^\perp = \mathscr{N}(\mathbf{A}')$$

and

$$\mathscr{N}(\mathbf{A})^\perp = \mathscr{R}(\mathbf{A}'),$$

i.e., the orthogonal complement of the range of $\mathbf{A}$ coincides with the null space of $\mathbf{A}'$ and the orthogonal complement of the null space of $\mathbf{A}$ coincides with the range of $\mathbf{A}'$.

PROOF:

In order to establish the first assertion, we must show that a vector $\mathbf{z}$ belongs to $\mathscr{R}(\mathbf{A})^\perp$ iff it belongs to $\mathscr{N}(\mathbf{A}')$. Suppose first that $\mathbf{z}$ belongs to $\mathscr{R}(\mathbf{A})^\perp$. This means, by the definition of the orthogonal complement, that

$$(\mathbf{y}, \mathbf{z}) = 0 \quad \text{for every} \quad \mathbf{y} \in \mathscr{R}(\mathbf{A}),$$

or, using the other notation for the inner product, that

$$\mathbf{y}'\mathbf{z} = 0 \quad \text{for every} \quad \mathbf{y} \in \mathscr{R}(\mathbf{A}).$$

Now, by the definition of $\mathscr{R}(\mathbf{A})$, this last statement is equivalent to

$$(\mathbf{Ax})'\mathbf{z} = 0 \quad \text{for every} \quad \mathbf{x} \in \mathbf{R}^n,$$

and, in view of Proposition 9.1, part (iv), one obtains

$$\mathbf{x}'(\mathbf{A}'\mathbf{z}) = 0 \quad \text{for every} \quad \mathbf{x} \in \mathbf{R}^n.$$

Now a statement of the form $\mathbf{x}'\mathbf{w} = 0$ for every $\mathbf{x} \in \mathbf{R}^n$ can hold iff $\mathbf{w} = \mathbf{0}$. This follows, for instance, from Proposition 7.2. Thus, we must have

$$\mathbf{A}'\mathbf{z} = \mathbf{0}$$

and this means that $\mathbf{z} \in \mathcal{N}(\mathbf{A}')$. Conversely, suppose that $\mathbf{z}$ belongs to $\mathcal{N}(\mathbf{A}')$. This means that $\mathbf{A}'\mathbf{z} = \mathbf{0}$. Taking the inner product of both sides with an arbitrary element $\mathbf{x}$ of $\mathbf{R}^n$, we obtain

$$\mathbf{x}'(\mathbf{A}'\mathbf{z}) = \mathbf{x}'\mathbf{0} = 0$$

from which it follows, by part (iv) of Proposition 9.1, that

$$(\mathbf{A}\mathbf{x})'\mathbf{z} = 0.$$

Hence, $\mathbf{z} \in \mathcal{R}(\mathbf{A})^{\perp}$ as needed. The second part of the proposition now follows immediately, by applying the first part to $\mathbf{A}'$ in place of $\mathbf{A}$ and making use of Proposition 7.8. ∎

Corollary

If $\mathbf{A}$ is a real $m \times n$ matrix, then

$$rank \ \mathbf{A} = rank \ \mathbf{A}'.$$

PROOF:
Let *rank* $\mathbf{A} = k$. By Proposition 7.6,

$$dim \ \mathcal{R}(\mathbf{A})^{\perp} = m - k.$$

But $\mathcal{R}(\mathbf{A})^{\perp} = \mathcal{N}(\mathbf{A}')$ and therefore

$$dim \ \mathcal{N}(\mathbf{A}') = m - k.$$

Now, by Proposition 8.1, $dim \ \mathcal{R}(\mathbf{A}') + dim \ \mathcal{N}(\mathbf{A}') = m$, and hence $dim \ \mathcal{R}(\mathbf{A}') = k$, as asserted. ∎

Once again, let $\mathbf{A}$ be a real $m \times n$ matrix. Recall that the range of $\mathbf{A}$ is the subspace spanned by its columns and that the rank of $\mathbf{A}$ is the dimension of this subspace. It is therefore obvious that the rank of $\mathbf{A}$ is equal to the maximal number of linearly independent columns of $\mathbf{A}$. In other words, *rank* $\mathbf{A} = k$ iff $\mathbf{A}$ has k linearly independent columns but does not have $k + 1$ linearly independent columns. Thus, the foregoing corollary has the following important implication: If $\mathbf{A}$ is any real matrix, then the maximal number of linearly independent columns of $\mathbf{A}$ is equal to the maximal number of linearly independent rows of $\mathbf{A}$.

9.1. Consider the following real 2×2 matrices, all denoted **A**:

(i) $\mathbf{A} = \begin{bmatrix} 1 & 1 \\ -1 & -1 \end{bmatrix}$;

(ii) $\mathbf{A} = \begin{bmatrix} 5 & 0 \\ 3 & 0 \end{bmatrix}$;

(iii) $\mathbf{A} = \begin{bmatrix} 1 & -2 \\ -3 & 6 \end{bmatrix}$;

(iv) $\mathbf{A} = \begin{bmatrix} 1 & 2 \\ 3 & 4 \end{bmatrix}$.

In each case, find $\mathscr{R}(\mathbf{A})$, $\mathscr{N}(\mathbf{A})$, $\mathscr{R}(\mathbf{A}')$, and $\mathscr{N}(\mathbf{A}')$, and exhibit all of them in a diagram.

9.2. Prove the following assertion: If **A** is a real $m \times n$ matrix, then

$$rank \ \mathbf{A}'\mathbf{A} = rank \ \mathbf{A}\mathbf{A}' = rank \ \mathbf{A}.$$

[Hint: Start by looking at $\mathscr{N}(\mathbf{A}'\mathbf{A})$ and $\mathscr{N}(\mathbf{A}\mathbf{A}')$.]

9.3. Let **A** be a real $m \times n$ matrix. Assume that $n < m$ and that $rank \ \mathbf{A} = n$. Now let **y** be a member of $\mathbf{R}^m$ that does not belong to $\mathscr{R}(\mathbf{A})$. Find a member $\hat{\mathbf{y}}$ of $\mathscr{R}(\mathbf{A})$ such that the inner product $(\mathbf{y} - \hat{\mathbf{y}}, \mathbf{y} - \hat{\mathbf{y}})$ satisfies the inequality

$$(\mathbf{y} - \hat{\mathbf{y}}, \mathbf{y} - \hat{\mathbf{y}}) \leq (\mathbf{y} - \mathbf{z}, \mathbf{y} - \mathbf{z}) \quad \text{for every} \quad \mathbf{z} \in \mathscr{R}(\mathbf{A}).$$

In other words, find a vector $\hat{\mathbf{x}} \in \mathbf{R}^n$ such that

$$(\mathbf{y} - \mathbf{A}\hat{\mathbf{x}}, \mathbf{y} - \mathbf{A}\hat{\mathbf{x}}) \leq (\mathbf{y} - \mathbf{A}\mathbf{x}, \mathbf{y} - \mathbf{A}\mathbf{x}) \quad \text{for every} \quad \mathbf{x} \in \mathbf{R}^n.$$

This problem occurs in statistics, where it goes by the name of *least squares estimation*. [Hint: Use Problem 7.6 on page 67.]

10 Square Matrices

Consider a real linear space L, and suppose that a new binary operation, called *multiplication*, is defined in L. Let the element of L that is associated with the pair $\langle x, y \rangle$ by multiplication be denoted xy, for every $x \in L, y \in L$. Now assume that multiplication satisfies the following conditions:

1. For every $x \in L, y \in L$, and $z \in L$,
   $$x(yz) = (xy)z.$$
2. For every $x \in L, y \in L$, and $z \in L$,
   $$x(y + z) = xy + xz.$$
3. For every $x \in L, y \in L$, and $z \in L$,
   $$(x + y)z = xz + yz.$$
4. For every $x \in L, y \in L$, and $\lambda \in \mathbf{R}$,
   $$\lambda(xy) = (\lambda x)y = x(\lambda y).$$

Condition 1 states that multiplication is associative. Conditions 2 and 3 govern the interaction of multiplication with addition. They state that multiplication distributes over addition both on the left and on the right. Condition 4 governs the interaction of multiplication with multiplication by a real number.

If L is a real linear space in which multiplication is defined, and if conditions 1–4 are satisfied, then L is referred to as a **real linear algebra**. If, furthermore, multiplication is commutative in L, then L is referred to as a **commutative** real linear algebra.

The most common example of a real linear algebra is the real number system, **R**. In this algebra, multiplication and multiplication by a real number amount to the same thing, but this does not prevent both from being well defined. **R** is, of course, a commutative algebra.

The set of all real $n \times n$ matrices, with addition and multiplication by a real number defined in the natural way and with matrix multiplication defined as in Chapter 8, is also a real linear algebra. To prove this, one must verify that conditions 1–4 above are satisfied, and this can be done directly, using the various definitions involved. The algebra of real $n \times n$ matrices is not commutative, for it is not true that $\mathbf{AB} = \mathbf{BA}$ for every real $n \times n$ matrix $\mathbf{A}$ and for every real $n \times n$ matrix $\mathbf{B}$.

A real $n \times n$ matrix is often referred to as a **real square matrix of order** $\boldsymbol{n}$. We shall use this term on some occasions, but in most cases we shall revert to the shorter "real $n \times n$ matrix." In the remainder of this chapter, we shall take a look at some properties of algebras of square matrices. Often, in proving an assertion about square matrices, it is useful to look at them as the natural matrices of linear transformations on $\mathbf{R}^n$ to $\mathbf{R}^n$, and to prove the assertions in terms of these linear transformations. This procedure is valid because the set of all linear transformations on $\mathbf{R}^n$ to $\mathbf{R}^n$ is also a real linear algebra,† and the function that associates with each linear transformation its natural matrix is an "algebra-isomorphism."

It is perhaps useful to look at some properties of the real number system, $\mathbf{R}$, and see whether or not we can find corresponding properties for the algebra of real $n \times n$ matrices. To do this, let us refer to the axioms for the real number system in Chapter 4. The first four axioms appear also in the definition of a real linear space, and therefore the algebra of real $n \times n$ matrices certainly satisfies them. The fifth axiom, associativity of multiplication, appears as condition 1 in the definition of a real linear algebra, but the sixth axiom, commutativity of multiplication, holds only in a commutative linear algebra. This brings us to the seventh axiom, which postulates the existence of an element, denoted 1, having the property that for every real number λ, $\lambda 1 = 1\lambda = \lambda$. Is there a similar element in the algebra of real $n \times n$ matrices? The answer is in the affirmative. Consider the identity matrix $\mathbf{I} = \langle \mathbf{e}_1, \mathbf{e}_2, \ldots, \mathbf{e}_n \rangle$, and recall that $\mathbf{I}$ is the natural matrix of the identity transformation on $\mathbf{R}^n$ onto $\mathbf{R}^n$. We know that composing any function with the identity (whether on the left or on the right) leaves the function unaltered, provided the composition is well defined. In terms of matrices, this assertion may be stated as follows: $\mathbf{AI} = \mathbf{IA} = \mathbf{A}$ for every real $n \times n$ matrix $\mathbf{A}$. This property can, of course, be verified directly, using the definition of matrix multiplication. Thus, we see that, in the algebra of real $n \times n$ matrices, the identity matrix $\mathbf{I}$ plays the same role that 1 plays in the real number system. Indeed, it is easy to show that, like the number 1 in the real number system, the matrix $\mathbf{I}$ is the *only* element of the algebra possessing this property. The proof of this uniqueness assertion is identical with the proof of the corresponding uniqueness assertion for the real number system.

We turn now to the eighth axiom for the real number system, which postulates that for each real number λ, except 0, there is a real number λ^{-1} such that $\lambda\lambda^{-1} = \lambda^{-1}\lambda = 1$. We have already seen in Chapter 8 that if T is an invertible linear transformation on $\mathbf{R}^n$ onto $\mathbf{R}^n$, and if $\mathbf{A}$ is the natural matrix of T, then there exists a real $n \times n$ matrix $\mathbf{A}^{-1}$, namely the natural matrix of T^{-1}, such that $\mathbf{AA}^{-1} = \mathbf{A}^{-1}\mathbf{A} = \mathbf{I}$. The following proposition says that the converse of this assertion is also true.

Proposition 10.1

Let T be a linear transformation on $\mathbf{R}^n$ to $\mathbf{R}^n$, and let $\mathbf{A}$ be the natural matrix of T. If T is not invertible, then there does not exist a real $n \times n$ matrix $\mathbf{B}$ with

---

† With addition and multiplication by a real number defined as in Problem 8.2, page 76, and with multiplication defined to coincide with composition.

the property that $\mathbf{AB} = \mathbf{I}$, *nor does there exist a real* $n \times n$ *matrix* $\mathbf{C}$ *with the property that* $\mathbf{CA} = \mathbf{I}$.

PROOF:

Suppose first that $\mathbf{AB} = \mathbf{I}$ for some real $n \times n$ matrix B, and let Q be the linear transformation whose natural matrix is $\mathbf{B}$. Then, the composition $T \circ Q$ must coincide with the identity, i.e., it must be true that

$$(T \circ Q)(\mathbf{x}) = \mathbf{x} \quad \text{for every} \quad \mathbf{x} \in \mathbf{R}^n.$$

But T is not invertible, and it follows from this that it cannot be onto. (How?) Let us select an element $\mathbf{x}$ of $\mathbf{R}^n$ which is not in the range of T. Then, we have $(T \circ Q)(\mathbf{x}) = T(Q(\mathbf{x}))$, and $T(Q(\mathbf{x}))$ cannot be equal to $\mathbf{x}$ no matter what $Q(\mathbf{x})$ is. Now assume that $\mathbf{CA} = \mathbf{I}$ for some real $n \times n$ matrix $\mathbf{C}$, and let P be the linear transformation whose natural matrix is $\mathbf{C}$. Then,

$$(P \circ T)(\mathbf{x}) = \mathbf{x} \quad \text{for every} \quad \mathbf{x} \in \mathbf{R}^n.$$

T is not invertible, and therefore it has a nontrivial null space. Select an element $\mathbf{x}$ of $\mathbf{R}^n$ such that $\mathbf{x} \neq \mathbf{0}$ and $T(\mathbf{x}) = \mathbf{0}$. By applying P to both sides of $T(\mathbf{x}) = \mathbf{0}$, we get

$$(P \circ T)(\mathbf{x}) = P(T(\mathbf{x})) = P(\mathbf{0}) = \mathbf{0},$$

which contradicts the previous assertion. ∎

Let us recapitulate: If $\mathbf{A}$, a real $n \times n$ matrix, is the natural matrix of an invertible linear transformation, then there exist real $n \times n$ matrices $\mathbf{B}$ and $\mathbf{C}$ such that $\mathbf{AB} = \mathbf{I}$ and $\mathbf{CA} = \mathbf{I}$. Furthermore, $\mathbf{B}$ and $\mathbf{C}$ must be equal to each other. (Proof: $(\mathbf{CA})\mathbf{B} = \mathbf{C}(\mathbf{AB})$ by the associativity of matrix multiplication. Hence, $\mathbf{IB} = \mathbf{CI}$ and, by the properties of the matrix $\mathbf{I}$, $\mathbf{B} = \mathbf{C}$.) Both matrices must, in fact, be equal to the matrix $\mathbf{A}^{-1}$, which is the natural matrix of the relevant inverse transformation. On the other hand, if $\mathbf{A}$ is the natural matrix of a noninvertible linear transformation, then neither a matrix $\mathbf{B}$ satisfying $\mathbf{AB} = \mathbf{I}$, nor a matrix $\mathbf{C}$ satisfying $\mathbf{CA} = \mathbf{I}$, exists.

Let A be a real $n \times n$ matrix. If the matrix $\mathbf{A}^{-1}$ exists, then $\mathbf{A}$ is said to be a **nonsingular** matrix, and $\mathbf{A}^{-1}$ is referred to as the **inverse** of $\mathbf{A}$. Otherwise, $\mathbf{A}$ is said to be a **singular** matrix. It follows from the discussion at the end of Chapter 9 that $\mathbf{A}$ is nonsingular iff its columns (or, equivalently, its rows) form a linearly independent set.

We now return to the axioms for the real number system. The eighth axiom postulates that every real number, except 0, has a reciprocal. In the algebra of real $n \times n$ matrices, the element corresponding to 0 is clearly the $n \times n$ matrix whose entries are all 0, for this is the matrix that, when added to another matrix, leaves that matrix unaltered. We refer to this matrix as the **null matrix**. The null matrix is clearly singular, and in this sense we have another instance of similarity between the algebra of real $n \times n$ matrices and the real number system. However, there are many nonnull singular matrices in the algebra of real $n \times n$ matrices (except when $n = 1$), namely, all the matrices whose columns (rows) are linearly dependent. Thus, with regard to the eighth axiom, the similarity between the real number system and the algebra of real $n \times n$ matrices is not complete.

The ninth axiom for the real number system requires that multiplication be distributive over addition. This is also true in the algebra of real $n \times n$ matrices (conditions 3 and 4). We see, then, that the algebra of real $n \times n$ matrices falls short of being a *field*, in that multiplication is not commutative and in that not all nonnull matrices have inverses. As for Axioms 10–14 in the definition of the real number system, they certainly do not hold in the algebra of real $n \times n$ matrices. (In some contexts, mathematicians do, in fact, define the relation "greater than . . ." for real $n \times n$ matrices, but this relation is not a complete ordering.)

Proposition 10.2

Let **A** *and* **B** *be real* $n \times n$ *matrices, and assume that both* **A** *and* **B** *are nonsingular. Then, the following assertions hold:*

(i) **A**′ *is nonsingular, and* $(\mathbf{A}')^{-1} = (\mathbf{A}^{-1})'$;
(ii) **AB** *is nonsingular, and* $(\mathbf{AB})^{-1} = \mathbf{B}^{-1}\mathbf{A}^{-1}$.

PROOF:

That **A**′ is nonsingular is immediate. To see that $(\mathbf{A}')^{-1} = (\mathbf{A}^{-1})'$, write $\mathbf{AA}^{-1} = \mathbf{I}$ and transpose both sides. The result is $(\mathbf{A}^{-1})'\mathbf{A}' = \mathbf{I}'$. But $\mathbf{I}' = \mathbf{I}$, so that we have $(\mathbf{A}^{-1})'\mathbf{A}' = \mathbf{I}$, which implies that $(\mathbf{A}^{-1})'$ is the inverse of **A**′. The fact that **AB** is nonsingular follows from the invertibility of the composition of two invertible functions. It is therefore possible to write $\mathbf{AB}(\mathbf{AB})^{-1} = \mathbf{I}$. Premultiplying by $\mathbf{B}^{-1}\mathbf{A}^{-1}$, we get the desired result. ∎

Let **A** and **B** be two real $n \times n$ matrices. We say that **A** is **similar** to **B** iff there exists a real $n \times n$ matrix **C** such that **C** is nonsingular and $\mathbf{B} = \mathbf{C}^{-1}\mathbf{AC}$.

Two things should be mentioned concerning the notion of similar matrices. First, similarity is an equivalence relation. Second, if **A** is the natural matrix of some linear transformation T on $\mathbf{R}^n$ to $\mathbf{R}^n$, then **B** is similar to

A iff **B** is the matrix of T relative to some basis for $\mathbf{R}^n$. Specifically, we have the following proposition:

Proposition 10.3

Consider a linear transformation T on $\mathbf{R}^n$ to $\mathbf{R}^n$, and let $\mathbf{A}$ and $\mathbf{C}$ be two real $n \times n$ matrices. Assume that $\mathbf{C}$ is nonsingular, so that the columns of $\mathbf{C}$ form a basis for $\mathbf{R}^n$. Then, $\mathbf{A}$ is the natural matrix of T iff $\mathbf{C}^{-1}\mathbf{AC}$ is the matrix of T relative to (the columns of) $\mathbf{C}$.†

PROOF:

Recall that if $\mathbf{x}$ is a member of $\mathbf{R}^n$, and if $\mathbf{w}$ is the n-tuple of coordinates of $\mathbf{x}$ relative to (the columns of) $\mathbf{C}$, then the relationship between $\mathbf{x}$ and $\mathbf{w}$ is given by

$$\mathbf{Cw} = \mathbf{x}$$

or by

$$\mathbf{w} = \mathbf{C}^{-1}\mathbf{x}.$$

Now assume that $\mathbf{A}$ is the natural matrix of T. The images of the columns of $\mathbf{C}$ under T are simply the columns of the matrix $\mathbf{AC}$. Therefore, to get the n-tuples of coordinates of these images relative to the basis $\mathbf{C}$, we have to premultiply $\mathbf{AC}$ by $\mathbf{C}^{-1}$. This leads to the matrix $\mathbf{C}^{-1}\mathbf{AC}$, as needed. Conversely, if $\mathbf{C}^{-1}\mathbf{AC}$ is the matrix of T relative to the basis $\mathbf{C}$, then the images of the columns of $\mathbf{C}$ under T are the columns of $\mathbf{AC}$. Now let $\mathbf{x}$ be an arbitrary member of $\mathbf{R}^n$, and let $\mathbf{w}$ be the n-tuple of coordinates of $\mathbf{x}$ relative to $\mathbf{C}$. Then, $\mathbf{x} = \mathbf{Cw}$ and $T(\mathbf{x}) = T(\mathbf{Cw}) = (\mathbf{AC})\mathbf{w} = \mathbf{A}(\mathbf{Cw}) = \mathbf{Ax}$, which proves the assertion. ∎

In the remainder of this chapter, we introduce several special types of real $n \times n$ matrices. These matrices will be of interest in later chapters.

A real $n \times n$ matrix $\mathbf{A}$ is said to be **symmetric** iff $\mathbf{A}' = \mathbf{A}$. It is said to be **skew-symmetric** iff $\mathbf{A}' = -\mathbf{A}$. We note in passing that every real $n \times n$ matrix is the sum of a symmetric matrix and a skew-symmetric matrix. (Proof: $\mathbf{A} = \frac{1}{2}(\mathbf{A} + \mathbf{A}') + \frac{1}{2}(\mathbf{A} - \mathbf{A}')$. The first member of the right-hand side is a symmetric matrix, while the second member is a skew-symmetric matrix.) If the entries of a real $n \times n$ matrix $\mathbf{A}$ are denoted a_{ij} (for $i = 1, 2, \dots, n$ and $j = 1, 2, \dots, n$) then the n-tuple $\langle a_{11}, a_{22}, \dots, a_{nn}\rangle$ is referred to as the **main diagonal** of $\mathbf{A}$. A real $n \times n$ matrix is said to be a **diagonal matrix** iff all the entries not in the main diagonal are 0. In other words,

† In defining the matrix of a linear transformation T on $\mathbf{R}^n$ to $\mathbf{R}^m$, we had to choose *two* bases, one for $\mathbf{R}^n$ and one for $\mathbf{R}^m$. Here, $m = n$, and the matrix $\mathbf{C}$ assumes the role of both bases.

$\mathbf{A} = [a_{ij}]$ is a diagonal matrix iff there exist n real numbers, $\lambda_1, \lambda_2, \ldots, \lambda_n$, such that $a_{ij} = \lambda_i \delta_{ij}$, where δ_{ij} is the Kronecker delta. A diagonal matrix is always a symmetric matrix, but a symmetric is clearly not, in general, a diagonal matrix. In Chapter 14 we shall show, however, that every symmetric matrix is *similar* to a diagonal matrix. Finally, a real $n \times n$ matrix $\mathbf{A}$ is said to be an **orthogonal matrix** iff $\mathbf{A}'\mathbf{A} = \mathbf{I}$. Note that the assertion $\mathbf{A}'\mathbf{A} = \mathbf{I}$ is equivalent to the assertion that the columns (or rows) of $\mathbf{A}$ form an orthonormal basis for $\mathbf{R}^n$.

10.1. In the course of proving Proposition 10.1, we have made use of the following assertion: If T is a linear transformation on $\mathbf{R}^n$ *onto* $\mathbf{R}^n$, then T is invertible. Prove this assertion.

10.2. Show that the relation "is similar to . . ." is an equivalence relation between real $n \times n$ matrices.

10.3. We know that if a real number λ satisfies the equation $\lambda = \lambda^{-1}$, then $\lambda = 1$ or $\lambda = -1$. Let $\mathbf{A}$ be a real $n \times n$ matrix. Is it true that if $\mathbf{A} = \mathbf{A}^{-1}$ then $\mathbf{A} = \mathbf{I}$ or $\mathbf{A} = -\mathbf{I}$? Prove your assertion.

10.4. Let $\mathbf{A}$ be a real $n \times n$ matrix. Show that $\mathbf{A}$ is skew-symmetric iff $(\mathbf{x}, \mathbf{A}\mathbf{x}) = 0$ (or, in the alternative notation, $\mathbf{x}'\mathbf{A}\mathbf{x} = 0$) for every $\mathbf{x} \in \mathbf{R}^n$. [Hint for the "if" part: Try $\mathbf{x} = \mathbf{e}_i + \mathbf{e}_j$.]

10.5. Let $\mathbf{A}$ be a real $n \times n$ matrix. Show that $\mathbf{A}\mathbf{X} = \mathbf{X}\mathbf{A}$ for every real $n \times n$ matrix $\mathbf{X}$ iff $\mathbf{A}$ is given by $\mathbf{A} = \lambda\mathbf{I}$ for some real number λ. [Hint for the "only if" part: Let $\mathbf{E}_{ij}$ be the real $n \times n$ matrix whose ij-th entry is 1, and whose other entries are all 0. Then set $\mathbf{X} = \mathbf{E}_{ij}$.] *Note*: A real $n \times n$ matrix of the form $\lambda\mathbf{I}$, where λ is some real number, is referred to as a **scalar matrix**.

10.6. Once again, consider a real $n \times n$ matrix $\mathbf{A}$. Show that $\mathbf{A}$ is an orthogonal matrix iff

$$(\mathbf{A}\mathbf{x}, \mathbf{A}\mathbf{x}) = (\mathbf{x}, \mathbf{x}) \quad \text{for every} \quad \mathbf{x} \in \mathbf{R}^n.$$

11 Linear Equations

Let a real $m \times n$ matrix $\mathbf{A}$, with entries a_{ij}, and a vector $\mathbf{b} \in \mathbf{R}^m$, with components b_i, be given. Our task is to find a vector $\mathbf{x} \in \mathbf{R}^n$ such that

$$\mathbf{Ax} = \mathbf{b}.$$

(Sometimes, the task is to find *all* vectors $\mathbf{x}$ satisfying this equation.) Writing this equation out in full, we get

$$a_{11}x_1 + a_{12}x_2 + \cdots + a_{1n}x_n = b_1$$
$$a_{21}x_1 + a_{22}x_2 + \cdots + a_{2n}x_n = b_2$$
$$\vdots$$
$$a_{m1}x_1 + a_{m2}x_2 + \cdots + a_{mn}x_n = b_m,$$

which is the familiar form of a system of m linear equations in n unknowns. Now consider the linear transformation, call it T, whose natural matrix is $\mathbf{A}$. The set

$$\{\mathbf{x} \,|\, \mathbf{x} \in \mathbf{R}^n, \mathbf{Ax} = \mathbf{b}\}$$

is precisely the *inverse image* of the singleton $\{\mathbf{b}\}$ under T. In the present context, however, this set goes by another name. It is usually called the **set of solutions** of the equation $Ax = b$, and its elements are referred to, naturally, as **solutions** of the equation $Ax = b$. The equation $Ax = b$ is said to be **homogeneous** iff $b = 0$. Note that the set of solutions of the homogeneous equation $Ax = 0$ is simply the null space of A. If the equation $Ax = b$ is not homogeneous (that is, if $b \neq 0$), then we refer to $Ax = 0$ as the homogeneous equation **associated with** the equation $Ax = b$.

Proposition 11.1

Let A be a real $m \times n$ matrix, and let b be a member of R^m. Finally, let x^* be a particular solution of the equation $Ax = b$. Then, an element $\hat{x}$ of R^n is also a solution of $Ax = b$ iff $\hat{x}$ is of the form

$$\hat{x} = x^* + u$$

where u is a solution of the homogeneous equation associated with the equation $Ax = b$.

PROOF:

Left to the reader. ∎

Consider the linear equation $\mathbf{Ax} = \mathbf{b}$. If $\mathbf{b}$ does not belong to the range of $\mathbf{A}$, then this equation has no solution; if $\mathbf{b}$ does belong to the range of $\mathbf{A}$, then there are two possibilities, depending on the rank of $\mathbf{A}$. Specifically, if *rank* $\mathbf{A} = n$, that is, if $\mathbf{A}$ has a trivial null space, then, assuming that $\mathbf{b}$ is in the range of $\mathbf{A}$, the equation $\mathbf{Ax} = \mathbf{b}$ has exactly one solution. This is so because $\mathbf{A}$ has a trivial null space iff it is the natural matrix of a one-to-one linear transformation. On the other hand, if *rank* $\mathbf{A} < n$, then the equation $\mathbf{Ax} = \mathbf{b}$ has infinitely many solutions, again assuming that $\mathbf{b}$ belongs to the range of $\mathbf{A}$. This assertion follows immediately from Proposition 11.1 above.

The null element of the space $\mathbf{R}^m$ belongs to the range of every real $m \times n$ matrix. Therefore, the homogeneous equation $\mathbf{Ax} = \mathbf{0}$ always has at least one solution, namely $\mathbf{x} = \mathbf{0}$. This solution is known as the **trivial solution** of the homogeneous equation. It is clear that the homogeneous equation $\mathbf{Ax} = \mathbf{0}$, where $\mathbf{A}$ is an $m \times n$ matrix, has a nontrivial solution iff *rank* $\mathbf{A} < n$. Indeed, if *rank* $\mathbf{A} = k$, then the homogeneous equation $\mathbf{Ax} = \mathbf{0}$ has exactly $n - k$ linearly independent (and hence, in particular, nontrivial) solutions. These assertions are merely restatements of our old propositions on the null space of a linear transformation.

By appealing to the properties of the transpose of a real matrix (Chapter 9), one can derive a criterion that tells whether or not a given linear equation has a solution. Specifically, we have:

Proposition 11.2

Let $\mathbf{A}$ be a real $m \times n$ matrix, and let $\mathbf{b}$ belong to $\mathbf{R}^m$. Then, exactly one of the following two statements holds true:

    (i) *The equation $\mathbf{Ax} = \mathbf{b}$ has a solution.*

    (ii) *The homogeneous equation $\mathbf{A'y} = \mathbf{0}$ has a solution $\hat{\mathbf{y}} \in \mathbf{R}^m$ satisfying $(\hat{\mathbf{y}}, \mathbf{b}) \neq 0$.*

PROOF:

(i) is true iff $\mathbf{b} \in \mathscr{R}(\mathbf{A})$. By Propositions 7.8 and 9.2, we have $\mathscr{R}(\mathbf{A}) = (\mathscr{R}(\mathbf{A})^\perp)^\perp = \mathscr{N}(\mathbf{A}')^\perp$. Therefore, (i) is true iff $\mathbf{b} \in \mathscr{N}(\mathbf{A}')^\perp$. But $\mathbf{b} \in \mathscr{N}(\mathbf{A}')^\perp$ is precisely the assertion that (ii) is false. ∎

Proposition 11.2 is sometimes referred to as the *theorem of the alternative* for linear equations. The reason for this name is clear from the statement of the proposition. As we shall see in the next chapter, "theorems of the alternative" occur also in the investigation of linear inequalities. From a practical point of view, the criterion for the solvability of a linear equation, which appears in Proposition 11.2, is of very little use. However, there are certain theoretical discussions in which Proposition 11.2 is, nevertheless, useful.

So far, our discussion of linear equations has amounted to little more than giving new names to old ideas. We are now ready to embark upon the central task of this chapter, which is to find a general method for solving linear equations. Our program will be a simple one: First, we shall look at a class of linear equations whose solutions are obtained essentially by inspection. Then, we shall develop a procedure by which every linear equation may be converted into one of these solution-by-inspection equations.

A real $m \times n$ matrix is called an **echelon matrix** iff the following conditions hold:

(i) The first nonzero entry in each row is 1.

(ii) If the first nonzero entry of some row (which must be 1 by the previous condition) appears in column j, then column j has no other nonzero entries. This for $j = 1, 2, \ldots, n$.

(iii) If row i has p leading zeros and row k has q leading zeros, and if $q > p$, then $k > i$. (Note that because of condition (ii), two rows cannot have the same number of leading zeros.)

For example, the matrix

$$\begin{bmatrix} 1 & -2 & 0 & 4 & 1 \\ 0 & 0 & 1 & -3 & -5 \\ 0 & 0 & 0 & 0 & 0 \end{bmatrix}$$

is a 3×5 echelon matrix. The null matrix and the identity matrix are both, obviously, echelon matrices.

Now consider the linear equation $\mathbf{Ax} = \mathbf{b}$. If $\mathbf{A}$ happens to be an echelon matrix, then the solutions of this equation are easily found. Let us, once again, write out the equation in full:

$$a_{11}x_1 + a_{12}x_2 + \cdots + a_{1n}x_n = b_1$$
$$a_{21}x_1 + a_{22}x_2 + \cdots + a_{2n}x_n = b_2$$
$$\vdots$$
$$a_{m1}x_1 + a_{m2}x_2 + \cdots + a_{mn}x_n = b_m.$$

The matrix $\mathbf{A}$ may have some rows in which all the entries are 0. By the definition of an echelon matrix, all of these rows must be concentrated at the bottom of the matrix. Say the last k rows of $\mathbf{A}$ are rows with only zero entries. Then, by looking at the last k components of the vector $\mathbf{b}$, we can determine whether or not the equation $\mathbf{Ax} = \mathbf{b}$ has a solution. For, if one of these last k components is not zero, then the equation obviously has no solution, i.e., $\mathbf{b}$ is not in the range of $\mathbf{A}$. Conversely, if all of the last k components of $\mathbf{b}$

are zero, then the equation $\mathbf{Ax} = \mathbf{b}$ has at least one solution, i.e., $\mathbf{b}$ does belong to the range of $\mathbf{A}$. This is proved simply by exhibiting a solution. In order to be able to describe a solution with ease, let us introduce the following *ad hoc* terminology: If the *j*-th column of the matrix $\mathbf{A}$ has in it the first nonzero entry of some row, then let us refer to x_j (the *j*-th component of the sought-after vector $\mathbf{x}$) as a *bound unknown*. On the other hand, if the *j*-th column of $\mathbf{A}$ does not have in it the first nonzero entry of some row, then let us refer to x_j as a *free unknown*. This for $j = 1, 2, \ldots, n$. Looking at the equation $\mathbf{Ax} = \mathbf{b}$ as a system of *m* equations in *n* unknowns, we observe that each bound unknown appears in one, and only one, equation. More precisely, if x_j is a bound unknown, then there is one equation in the system in which x_j appears with the coefficient 1, and in all other equations it appears with the coefficient 0. This follows directly from the fact that $\mathbf{A}$ is an echelon matrix. It also follows from this fact that if the bound unknown x_j appears in equation *k* with the coefficient 1, then all the other bound unknowns enter equation *k* with the coefficient 0. In other words, each equation involves at most one bound unknown, together (possibly) with some free unknowns. To obtain a solution of the equation $\mathbf{Ax} = \mathbf{b}$, one begins by assigning arbitrary values to all the free unknowns. (For example, one might set all the free unknowns equal to 0.) Now, if x_j is a bound unknown, then the value of x_j is found in the following manner: We look for the one equation where x_j appears with the coefficient 1. Say this is equation *k*. Equation *k* will, in general, involve some free unknowns as well, but it will not involve any other bound unknowns. Now, to the free unknowns we have already assigned certain values, so that equation *k* reduces to the following (linear) equation in the single unknown x_j:

$$x_j + \lambda = b_k$$

where λ is a real number depending on the arbitrary values that have been assigned to the free unknowns. From $x_j + \lambda = b_k$ we go to $x_j = b_k - \lambda$ using the properties of the real number system, and doing this for all bound unknowns completes the solution. It is clear, furthermore, that *all* the solutions of the equation $\mathbf{Ax} = \mathbf{b}$ can be obtained in this manner, by varying the arbitrary assignment of values to the free unknowns.

Now let us consider the equation $\mathbf{Ax} = \mathbf{b}$, where the real $m \times n$ matrix $\mathbf{A}$ is no longer assumed to be an echelon matrix. We shall find it convenient to work with the real $m \times (n + 1)$ matrix whose first *n* columns are the columns of $\mathbf{A}$ and whose last column is the vector $\mathbf{b}$. This matrix will be referred to as the **augmented matrix** associated with the equation $\mathbf{Ax} = \mathbf{b}$, and it will be denoted $[\mathbf{A}, \mathbf{b}]$. Another definition that we shall need is the following: A real $m \times n$ matrix $\mathbf{A}$ is said to be **row-equivalent** to a real $m \times n$ matrix $\mathbf{B}$ iff there exists a real $m \times m$ matrix $\mathbf{C}$ such that $\mathbf{C}$ is nonsingular and $\mathbf{B} = \mathbf{CA}$. It is clear that row-equivalence is an equivalence relation, so that when a matrix $\mathbf{A}$ is row-equivalent to a matrix $\mathbf{B}$, we are entitled to say, more concisely, that $\mathbf{A}$ and $\mathbf{B}$ are row-equivalent.

Proposition 11.3

Let $\mathbf{A}_1$ and $\mathbf{A}_2$ be real $m \times n$ matrices, and let $\mathbf{b}_1$ and $\mathbf{b}_2$ be members of the space $\mathbf{R}^m$. If the augmented matrices $[\mathbf{A}_1, \mathbf{b}_1]$ and $[\mathbf{A}_2, \mathbf{b}_2]$ are row-equivalent, then the equations $\mathbf{A}_1\mathbf{x} = \mathbf{b}_1$ and $\mathbf{A}_2\mathbf{x} = \mathbf{b}_2$ have the same set of solutions.

PROOF:

Let $\mathbf{C}$ be a nonsingular $m \times m$ matrix such that $[\mathbf{A}_2, \mathbf{b}_2] = \mathbf{C}[\mathbf{A}_1, \mathbf{b}_1]$. Note that $\mathbf{C}[\mathbf{A}_1, \mathbf{b}_1] = [\mathbf{CA}_1, \mathbf{Cb}_1]$. Now assume that a vector $\hat{\mathbf{x}}$ in $\mathbf{R}^n$ has been found such that $\mathbf{A}_1\hat{\mathbf{x}} = \mathbf{b}_1$ (i.e., $\hat{\mathbf{x}}$ is a solution of $\mathbf{A}_1\mathbf{x} = \mathbf{b}_1$). Premultiplying by $\mathbf{C}$, we obtain $\mathbf{CA}_1\hat{\mathbf{x}} = \mathbf{Cb}_1$, or $\mathbf{A}_2\hat{\mathbf{x}} = \mathbf{b}_2$, which means that $\hat{\mathbf{x}}$ is also a solution of $\mathbf{A}_2\mathbf{x} = \mathbf{b}_2$. The very same argument, with $\mathbf{C}^{-1}$ replacing $\mathbf{C}$, shows that if $\hat{\mathbf{x}}$ is a solution of $\mathbf{A}_2\mathbf{x} = \mathbf{b}_2$ then it is also a solution of $\mathbf{A}_1\mathbf{x} = \mathbf{b}_1$. ∎

In view of this proposition, it is clear that if we wish to find the solutions of the equation $\mathbf{Ax} = \mathbf{b}$, we ought to consider the following course of action: Take the augmented matrix $[\mathbf{A}, \mathbf{b}]$ and try to find another augmented matrix, say $[\mathbf{A}^*, \mathbf{b}^*]$, such that $[\mathbf{A}, \mathbf{b}]$ and $[\mathbf{A}^*, \mathbf{b}^*]$ are row-equivalent and such that $\mathbf{A}^*$ is an echelon matrix. We know how to find the set of solutions of the equation $\mathbf{A}^*\mathbf{x} = \mathbf{b}^*$ and, by Proposition 11.3, this set coincides with the set of solutions of the equation $\mathbf{Ax} = \mathbf{b}$.

Consider the augmented matrix, $[\mathbf{A}, \mathbf{b}]$, which is associated with the equation $\mathbf{Ax} = \mathbf{b}$. We shall now proceed to modify this augmented matrix by manipulating its rows. In particular, we shall permit three types of operations, known as **elementary operations**, to be performed on the rows of $[\mathbf{A}, \mathbf{b}]$. An elementary operation of the *first* type consists of taking two rows, say row i and row k, and interchanging them. An elementary operation of the *second* type consists of taking, say, the i-th row of the matrix in question and replacing it by some nonzero multiple of itself. Thus, to perform an elementary operation of the second type on the matrix $[\mathbf{A}, \mathbf{b}]$, we take the row $\langle a_{i1}, a_{i2}, \ldots, a_{in}, b_i \rangle$ and replace it by the row $\langle \lambda a_{i1}, \lambda a_{i2}, \ldots, \lambda a_{in}, \lambda b_i \rangle$, where λ is some real number other than zero. Finally, an elementary operation of the *third* type consists of taking two different rows of the matrix in question, say row i and row k, and replacing row i by the *sum* of row i and some multiple of row k. For the matrix $[\mathbf{A}, \mathbf{b}]$, this would mean replacing the row $\langle a_{i1}, a_{i2}, \ldots, a_{in}, b_i \rangle$ by $\langle a_{i1} + \lambda a_{k1}, a_{i2} + \lambda a_{k2}, \ldots, a_{in} + \lambda a_{kn}, b_i + \lambda b_k \rangle$ where λ is some real number and $i \neq k$.

Let $\mathbf{I}$ be the $m \times m$ identity matrix. Suppose now that the i-th row and the k-th row of $\mathbf{I}$ are interchanged. Call the resulting matrix $\mathbf{E}_1$. Similarly, suppose that the i-th row of $\mathbf{I}$ is replaced by λ times the i-th row, where $\lambda \neq 0$. Call the resulting matrix $\mathbf{E}_2$. Finally, let $\mathbf{E}_3$ be the matrix obtained when the i-th row of $\mathbf{I}$ is replaced by the sum of the i-th row and λ times the k-th row. (In other words, $\mathbf{E}_3$ is the matrix $\mathbf{I}$ with the ik-th entry changed from 0 to λ, assuming $i \neq k$.) Matrices of the types of $\mathbf{E}_1$, $\mathbf{E}_2$, and $\mathbf{E}_3$ are called **elementary matrices**.

Proposition 11.4

The elementary matrices $\mathbf{E}_1$, $\mathbf{E}_2$, *and* $\mathbf{E}_3$ *are all nonsingular.*

PROOF:

We shall show that these matrices are nonsingular by exhibiting their inverses. The inverse of $\mathbf{E}_1$ is again $\mathbf{E}_1$. The inverse of $\mathbf{E}_2$ is an elementary matrix like $\mathbf{E}_2$, except that the i-th row of $\mathbf{I}$ is multiplied by $1/\lambda$ instead of λ. (Recall that for $\mathbf{E}_2$ we require $\lambda \neq 0$.) Finally, the inverse of $\mathbf{E}_3$ is again an elementary matrix of the same type as $\mathbf{E}_3$, namely the matrix obtained when the ik-th entry of $\mathbf{I}$ is changed from 0 to $-\lambda$. ∎

Proposition 11.5

Let $\mathbf{A}$ *be a real* $m \times n$ *matrix. Then, performing an elementary operation of the first, second, or third type on the rows of* $\mathbf{A}$ *is equivalent to premultiplying* $\mathbf{A}$ *by an elementary matrix of the type of* $\mathbf{E}_1$, $\mathbf{E}_2$, *or* $\mathbf{E}_3$, *respectively.*

PROOF:

Left to the reader. ∎

The upshot of the foregoing propositions is that if, by means of an elementary operation (or by means of a finite sequence of elementary operations), we can get from the matrix $\mathbf{A}$ to the matrix $\mathbf{B}$, then $\mathbf{A}$ and $\mathbf{B}$ are row-equivalent.

Returning to the equation $\mathbf{Ax} = \mathbf{b}$, where $\mathbf{A}$ is a real $m \times n$ matrix and $\mathbf{b}$ is a vector in $\mathbf{R}^m$, we must now show that the augmented matrix $[\mathbf{A}, \mathbf{b}]$ can be converted, by means of a finite sequence of elementary operations, into a matrix $[\mathbf{A}^*, \mathbf{b}^*]$, where $\mathbf{A}^*$ is an echelon matrix. Rather than do this for a general matrix $\mathbf{A}$ and a general vector $\mathbf{b}$, let us consider a specific example. The reader will have no trouble convincing himself that the procedure used to treat this specific example is a perfectly general procedure.

Consider the following system of three linear equations in four unknowns:

$$
\begin{aligned}
x_2 + 2x_3 + x_4 &= 0 \\
3x_1 + 2x_2 - 2x_3 \phantom{{}+ 3x_4} &= 2 \\
4x_1 \phantom{{}+ 2x_2} - x_3 - 3x_4 &= 3.
\end{aligned}
$$

The augmented matrix associated with this system is

$$\begin{bmatrix} 0 & 1 & 2 & 1 & 0 \\ 3 & 2 & -2 & 0 & 2 \\ 4 & 0 & -1 & -3 & 3 \end{bmatrix}.$$

The first elementary operation is to interchange the first row and the second row. This is necessary because the first entry in the first row is 0. The operation leads to the matrix

$$\begin{bmatrix} 3 & 2 & -2 & 0 & 2 \\ 0 & 1 & 2 & 1 & 0 \\ 4 & 0 & -1 & -3 & 3 \end{bmatrix}.$$

Now, multiply the first row by $\frac{1}{3}$, so that the first entry in the first row will be 1:

$$\begin{bmatrix} 1 & \frac{2}{3} & -\frac{2}{3} & 0 & \frac{2}{3} \\ 0 & 1 & 2 & 1 & 0 \\ 4 & 0 & -1 & -3 & 3 \end{bmatrix}.$$

Next, replace the third row by the sum of the third row and -4 times the first row:

$$\begin{bmatrix} 1 & \frac{2}{3} & -\frac{2}{3} & 0 & \frac{2}{3} \\ 0 & 1 & 2 & 1 & 0 \\ 0 & -\frac{8}{3} & \frac{5}{3} & -3 & \frac{1}{3} \end{bmatrix}.$$

So far, the operations have resulted in the reduction of the first column to the required form. We switch now to the second column. The second entry in this column is 1, so there is no need to replace the second row by a multiple of itself. We therefore proceed to replace the first row by the sum of the first row and $-\frac{2}{3}$ times the second row. This leads to

$$\begin{bmatrix} 1 & 0 & -2 & -\frac{2}{3} & \frac{2}{3} \\ 0 & 1 & 2 & 1 & 0 \\ 0 & -\frac{8}{3} & \frac{5}{3} & -3 & \frac{1}{3} \end{bmatrix}.$$

Replacing the third row by the sum of the third row and $\frac{8}{3}$ times the second row, we get

$$\begin{bmatrix} 1 & 0 & -2 & -\frac{2}{3} & \frac{2}{3} \\ 0 & 1 & 2 & 1 & 0 \\ 0 & 0 & 7 & -\frac{1}{3} & \frac{1}{3} \end{bmatrix}.$$

Moving now to operations affecting the third column, we begin by dividing the third row by 7 (i.e., replacing it by $\frac{1}{7}$ times itself):

$$\begin{bmatrix} 1 & 0 & -2 & -\frac{2}{3} & \frac{2}{3} \\ 0 & 1 & 2 & 1 & 0 \\ 0 & 0 & 1 & -\frac{1}{21} & \frac{1}{21} \end{bmatrix}.$$

Replacing the first row by the sum of the first row and twice the third row leads to

$$\begin{bmatrix} 1 & 0 & 0 & -\frac{16}{21} & \frac{16}{21} \\ 0 & 1 & 2 & 1 & 0 \\ 0 & 0 & 1 & -\frac{1}{21} & \frac{1}{21} \end{bmatrix}.$$

Finally, replacing the second row by the sum of the second row and -2 times the third row, we get

$$\begin{bmatrix} 1 & 0 & 0 & -\frac{16}{21} & \frac{16}{21} \\ 0 & 1 & 0 & \frac{23}{21} & -\frac{2}{21} \\ 0 & 0 & 1 & -\frac{1}{21} & \frac{1}{21} \end{bmatrix}.$$

This matrix is of the form $[\mathbf{A}^*, \mathbf{b}^*]$, where $\mathbf{A}^*$ is an echelon matrix. In other words, we have reached our destination. In the terminology introduced previously, we have here three bound unknowns, namely x_1, x_2, and x_3, and one free unknown, x_4. Recall that in order to obtain a solution, we must set the values of the free unknowns arbitrarily and then solve for the bound unknowns. Let us, then, set $x_4 = 0$. This leads to

$$x_1 = \tfrac{16}{21},\ x_2 = -\tfrac{2}{21},\ x_3 = \tfrac{1}{21},\ x_4 = 0$$

or to

$$\mathbf{x} = \begin{bmatrix} \frac{16}{21} \\ -\frac{2}{21} \\ \frac{1}{21} \\ 0 \end{bmatrix},$$

which is, indeed, a solution of the original system of equations. In fact, as we know, for each arbitrary assignment of the value of x_4, we get a solution of the original system. For example, if we set $x_4 = -1$, we get

$$\mathbf{x} = \begin{bmatrix} 0 \\ 1 \\ 0 \\ -1 \end{bmatrix}.$$

In order to describe *all* the solutions of our system of equations, let us take the roundabout way and resort to Proposition 11.1. Accordingly, we shall seek the solutions **u** of the homogeneous equation associated with our system of equations. Clearly, if the augmented matrix of the equation **Ax** = **b** is reduced by elementary operations to the form [**A***, **b***], then the augmented matrix of the corresponding homogeneous equation will be reduced by the *same* elementary operations to the form [**A***, **0**]. Therefore, we conclude that the augmented matrix of the homogeneous equation associated with our original equation is row-equivalent to the matrix

$$\begin{bmatrix} 1 & 0 & 0 & -\frac{16}{21} & 0 \\ 0 & 1 & 0 & \frac{23}{21} & 0 \\ 0 & 0 & 1 & -\frac{1}{21} & 0 \end{bmatrix},$$

from which we can get the solutions of the homogeneous equation. We set the value of the free unknown, u_4, and then solve for the other unknowns. However, since we are interested in all the solutions, let us set u_4 equal to an arbitrary real number, λ. This leads to the solution

$$\mathbf{u} = \begin{bmatrix} \frac{16}{21}\lambda \\ -\frac{23}{21}\lambda \\ \frac{1}{21}\lambda \\ \lambda \end{bmatrix}$$

for the homogeneous equation. Using Proposition 11.1, together with one of the particular solutions we already have for the original equation, we obtain the following characterization for the set of solutions: **x** is a solution of the original equation iff it is of the form

$$\mathbf{x} = \begin{bmatrix} \frac{16}{21}\lambda \\ 1 - \frac{23}{21}\lambda \\ \frac{1}{21}\lambda \\ \lambda - 1 \end{bmatrix}$$

for some real number λ. In practice, of course, there is no need to go through Proposition 11.1. The entire set of solutions may be characterized directly, by looking at the matrix [**A***, **b***] and setting the free unknowns equal to arbitrary, unspecified, real numbers.

As has already been mentioned, we solved this example by a completely general procedure, capable of handling any equation of the form **Ax** = **b**. We shall not describe the procedure in general, because such a description is bound to involve rather tedious notation and this is apt to be more confusing than revealing.

Proposition 11.6

Every real m × n matrix is row-equivalent to an echelon matrix.

PROOF:
Use the procedure of the foregoing example. ∎

Corollary

*Let **A** be a real n × n matrix, and assume that **A** is nonsingular. Then, by a finite sequence of elementary operations, **A** can be converted into the identity matrix, **I**.*

PROOF:
It follows from Propositions 11.4 and 11.5 that an elementary operation converts a nonsingular matrix into a nonsingular matrix. Now, the only $n \times n$ echelon matrix that is also a nonsingular matrix is the identity matrix. ∎

The importance of this last corollary is that it points to a useful way of computing the inverse of a nonsingular matrix. Recall that in performing a finite sequence of elementary operations on the rows of a matrix **A**, we are really premultiplying **A** by a sequence of elementary matrices. In other words, what we are really doing when we use elementary operations to reduce the matrix **A** to the identity matrix **I** is finding a sequence of elementary matrices, say, $\mathbf{E}^{(1)}, \mathbf{E}^{(2)}, \ldots, \mathbf{E}^{(k)}$, such that

$$\mathbf{E}^{(k)}\mathbf{E}^{(k-1)} \cdots \mathbf{E}^{(2)}\mathbf{E}^{(1)}\mathbf{A} = \mathbf{I}.$$

But this means that the product $\mathbf{E}^{(k)}\mathbf{E}^{(k-1)} \cdots \mathbf{E}^{(2)}\mathbf{E}^{(1)}$ is precisely the matrix $\mathbf{A}^{-1}$. Thus, by keeping track of the elementary matrices involved in the elementary operations, we can find the inverse of **A**. As an example, let us find the inverse of the following matrix:

$$\mathbf{A} = \begin{bmatrix} 2 & 2 & 1 \\ 0 & -2 & -1 \\ -3 & 0 & 3 \end{bmatrix}.$$

In order to keep track of the elementary matrices involved in the operations we are about to perform, let us repeat the matrix **A** on the left side of the page, and on its right side let us write down the identity matrix, **I**. The right-hand side of the page will be used, so to speak, to accumulate our elementary matrices.

$$\begin{bmatrix} 2 & 2 & 1 \\ 0 & -2 & -1 \\ -3 & 0 & 3 \end{bmatrix} \qquad \begin{bmatrix} 1 & 0 & 0 \\ 0 & 1 & 0 \\ 0 & 0 & 1 \end{bmatrix}.$$

We start by dividing the first row by 2, and we do it to *both* matrices. The result is:

$$\begin{bmatrix} 1 & 1 & \frac{1}{2} \\ 0 & -2 & -1 \\ -3 & 0 & 3 \end{bmatrix} \qquad \begin{bmatrix} \frac{1}{2} & 0 & 0 \\ 0 & 1 & 0 \\ 0 & 0 & 1 \end{bmatrix}.$$

Now, we replace the third row with the sum of the third row and 3 times the first row. Again, we do it for both matrices:

$$\begin{bmatrix} 1 & 1 & \frac{1}{2} \\ 0 & -2 & -1 \\ 0 & 3 & \frac{9}{2} \end{bmatrix} \qquad \begin{bmatrix} \frac{1}{2} & 0 & 0 \\ 0 & 1 & 0 \\ \frac{3}{2} & 0 & 1 \end{bmatrix}.$$

The next operation is one of dividing the second row by -2:

$$\begin{bmatrix} 1 & 1 & \frac{1}{2} \\ 0 & 1 & \frac{1}{2} \\ 0 & 3 & \frac{9}{2} \end{bmatrix} \qquad \begin{bmatrix} \frac{1}{2} & 0 & 0 \\ 0 & -\frac{1}{2} & 0 \\ \frac{3}{2} & 0 & 1 \end{bmatrix}.$$

Subtracting the second row from the first row yields:

$$\begin{bmatrix} 1 & 0 & 0 \\ 0 & 1 & \frac{1}{2} \\ 0 & 3 & \frac{9}{2} \end{bmatrix} \qquad \begin{bmatrix} \frac{1}{2} & \frac{1}{2} & 0 \\ 0 & -\frac{1}{2} & 0 \\ \frac{3}{2} & 0 & 1 \end{bmatrix}.$$

We now replace the third row with the sum of the third row and -3 times the second row:

$$\begin{bmatrix} 1 & 0 & 0 \\ 0 & 1 & \frac{1}{2} \\ 0 & 0 & 3 \end{bmatrix} \qquad \begin{bmatrix} \frac{1}{2} & \frac{1}{2} & 0 \\ 0 & -\frac{1}{2} & 0 \\ \frac{3}{2} & \frac{3}{2} & 1 \end{bmatrix}.$$

Dividing the third row by 3 leads to:

$$\begin{bmatrix} 1 & 0 & 0 \\ 0 & 1 & \frac{1}{2} \\ 0 & 0 & 1 \end{bmatrix} \qquad \begin{bmatrix} \frac{1}{2} & \frac{1}{2} & 0 \\ 0 & -\frac{1}{2} & 0 \\ \frac{1}{2} & \frac{1}{2} & \frac{1}{3} \end{bmatrix}.$$

Finally, to obtain the identity matrix on the left-hand side of the page, we replace the second row with the sum of the second row and $-\frac{1}{2}$ times the third row:

$$\begin{bmatrix} 1 & 0 & 0 \\ 0 & 1 & 0 \\ 0 & 0 & 1 \end{bmatrix} \qquad \begin{bmatrix} \frac{1}{2} & \frac{1}{2} & 0 \\ -\frac{1}{4} & -\frac{3}{4} & -\frac{1}{6} \\ \frac{1}{2} & \frac{1}{2} & \frac{1}{3} \end{bmatrix}.$$

The last matrix on the right-hand side of the page is $\mathbf{A}^{-1}$.

PROBLEMS

11.1.   Prove Proposition 11.1.

11.2.   Is the converse of Proposition 11.3 true?

11.3.   Consider the linear equation $\mathbf{Ax} = \mathbf{b}$, where $\mathbf{A}$ is a real $m \times n$ matrix and $\mathbf{b}$ is a vector in $\mathbf{R}^m$, and assume that it has a solution. If *rank* $\mathbf{A} = n$, then this solution is necessarily unique and, furthermore, there is a way to write the solution in closed form, entirely in matrix notation. Find this closed-form expression and compare it with the answer to Problem 9.3 on page 83.

11.4.   Show that every nonsingular $n \times n$ matrix is the product of a finite number of elementary matrices.

11.5.   Find the complete set of solutions of the equation $\mathbf{Ax} = \mathbf{b}$ for each of the following specifications of $\mathbf{A}$ and $\mathbf{b}$:

$$\text{(i)}\quad \mathbf{A} = \begin{bmatrix} 1 & 1 & 1 & 1 \\ 1 & 2 & 0 & 1 \\ -3 & -2 & -1 & -2 \\ 3 & 0 & -3 & 0 \end{bmatrix} \quad \mathbf{b} = \begin{bmatrix} -1 \\ -1 \\ 2 \\ 0 \end{bmatrix};$$

$$\text{(ii)}\quad \mathbf{A} = \begin{bmatrix} -2 & -3 & 11 & -14 \\ 1 & -1 & 2 & 2 \\ 3 & -1 & 0 & 10 \end{bmatrix} \quad \mathbf{b} = \begin{bmatrix} 3 \\ -3 \\ -3 \end{bmatrix};$$

$$\text{(iii)}\quad \mathbf{A} = \begin{bmatrix} 3 & 6 & 1 & -5 & 6 \\ 0 & 0 & -5 & 3 & 5 \\ 1 & -3 & 2 & -1 & -4 \end{bmatrix} \quad \mathbf{b} = \begin{bmatrix} 10 \\ 0 \\ -1 \end{bmatrix}.$$

11.6. Using elementary operations, find the inverses of the following three matrices:

(i)
$$\begin{bmatrix} 1 & 0 & -2 & 0 \\ -\frac{1}{3} & 0 & 0 & -\frac{2}{3} \\ 0 & -1 & \frac{1}{3} & \frac{1}{3} \\ -1 & \frac{2}{3} & 0 & 0 \end{bmatrix};$$

(ii)
$$\begin{bmatrix} \alpha & 0 & \beta \\ 0 & \gamma & 0 \\ \beta & 0 & \alpha \end{bmatrix};$$

(iii)
$$\begin{bmatrix} 1 & \lambda & 0 & 0 & 0 \\ \lambda & 1 & \lambda & 0 & 0 \\ 0 & \lambda & 1 & \lambda & 0 \\ 0 & 0 & \lambda & 1 & \lambda \\ 0 & 0 & 0 & \lambda & 1 \end{bmatrix}.$$

12 Linear Inequalities

In the social sciences, it often happens that the behavior of a system is described by inequalities rather than by equations (or, sometimes, by inequalities *in addition* to equations). For example, in many cases the variables being studied are restricted, by their very nature, to assume only nonnegative values. These nonnegativity restrictions take the form of inequalities. It is perhaps not an exaggeration to say that familiarity with inequalities (and with linear inequalities in particular) is indispensable in the social sciences. At any rate, a book on linear algebra for social sciences would be incomplete without a discussion of linear inequalities. It is not a disregard for the needs of the social scientist that prevents most books on linear algebra from going into the theory of linear inequalities. The reason lies rather in the fact that the theory of linear inequalities is really a branch of the study of convex sets, and the study of convex sets goes beyond linear algebra. In this chapter, we shall make an attempt to present some of the basic results in the theory of linear inequalities without explicit reference to the study of convex sets. At one point, in the course of proving Proposition 12.1, we shall resort to an argument that is nonalgebraic in nature (it has to do with such notions as convergence of a sequence), and this argument will therefore be given in an appendix, at the end of the chapter.

Let us begin by introducing some new terminology and notation. Consider an element, **x**, of $\mathbf{R}^n$, with components $x_1, \ldots, x_n$. We shall say that **x** is **nonnegative** iff x_i is a nonnegative real number for $i = 1, \ldots, n$. We shall say that **x** is **semipositive** iff **x** is nonnegative and different from **0**. Finally, we shall say that **x** is **positive** iff x_i is a positive real number for $i = 1, \ldots, n$. As for notation, let us agree to write $\mathbf{x} \geq \mathbf{0}$ iff **x** is nonnegative, $\mathbf{x} > \mathbf{0}$ iff **x** is semipositive, and $\mathbf{x} \gg \mathbf{0}$ iff **x** is positive. If **x** and **y** are members of $\mathbf{R}^n$, then we shall write

$$\mathbf{x} \geq \mathbf{y}, \quad \mathbf{x} > \mathbf{y}, \quad \mathbf{x} \gg \mathbf{y}$$

according as

$$\mathbf{x} - \mathbf{y} \geq \mathbf{0}, \quad \mathbf{x} - \mathbf{y} > \mathbf{0}, \quad \mathbf{x} - \mathbf{y} \gg \mathbf{0}.$$

Proposition 12.1

Let M be a subspace of $\mathbf{R}^n$, and assume that there are no semipositive vectors in M. Let dim $M = k$. If $k < n - 1$, then M can be embedded in another subspace, of dimension $k + 1$, which also does not possess a semipositive vector. (In other words, if $k < n - 1$, then there exists a subspace V such that $V \supset M$, dim $V = k + 1$, and such that there are no semipositive vectors in V.)

107

PROOF:

Let L be a subspace of $\mathbf{R}^n$, with the following properties: $L \supset M$ and $\dim L = k + 2$. We shall now proceed to find a $(k + 1)$-dimensional subspace of L that still contains M, and that has no semipositive vectors. If L itself has no semipositive vectors, then the existence of such a subspace is immediate. Therefore, assume that L *does* possess a semipositive vector. Let $\mathbf{B} = \langle \mathbf{x}_1, \ldots, \mathbf{x}_{k+2} \rangle$ be a basis for L, with the property that $\langle \mathbf{x}_1, \ldots, \mathbf{x}_k \rangle$ is a basis for M. Define a set A in the following manner:

$$A = \left\{ \langle \alpha_{k+1}, \alpha_{k+2} \rangle \middle| \begin{array}{l} \textit{there exists a k-tuple } \langle \alpha_1, \ldots, \alpha_k \rangle \\ \textit{such that } \sum_{i=1}^{k+2} \alpha_i \mathbf{x}_i \geqq \mathbf{0} \end{array} \right\}.$$

Thus, a pair $\langle \alpha_{k+1}, \alpha_{k+2} \rangle$ belongs to A iff there exists a k-tuple $\langle \alpha_1, \ldots, \alpha_k \rangle$ such that $\langle \alpha_1, \ldots, \alpha_{k+2} \rangle$ is the $(k + 2)$-tuple of coordinates of a nonnegative vector, relative to the basis $\mathbf{B}$. Since A is a set of pairs of real numbers, it will be convenient to think of it graphically, as a set of points in the plane.

We claim that if there exists in the plane a straight line whose only point in common with the set A is the origin, $\langle 0, 0 \rangle$, then our proposition is proved. To see this, let D be a straight line through the origin in the plane. Let $\langle \beta, \gamma \rangle$ be an arbitrary point on the line D, other than the origin. Then, D is given by

$$D = \{ \langle \lambda\beta, \lambda\gamma \rangle \mid \lambda \in \mathbf{R} \}.$$

Consider the subspace of $\mathbf{R}^n$ that is spanned by the $(k + 1)$-tuple $\langle \mathbf{x}_1, \mathbf{x}_2, \ldots, \mathbf{x}_k, \beta\mathbf{x}_{k+1} + \gamma\mathbf{x}_{k+2} \rangle$. This subspace, call it V, certainly contains M, and its dimension is $k + 1$. Now, if $D \cap A = \{ \langle 0, 0 \rangle \}$, then V does not possess a semipositive vector, for if it did, we would have a $(k + 1)$-tuple of real numbers, $\langle \alpha_1, \ldots, \alpha_k, \alpha_{k+1} \rangle$, such that

$$\sum_{i=1}^{k} \alpha_i \mathbf{x}_i + \beta\alpha_{k+1}\mathbf{x}_{k+1} + \gamma\alpha_{k+1}\mathbf{x}_{k+2} > \mathbf{0}.$$

Note that α_{k+1} cannot be zero. This follows from the foregoing inequality, taken together with the fact that M does not possess a semipositive vector. But, by construction, the point $\langle \beta\alpha_{k+1}, \gamma\alpha_{k+1} \rangle$ belongs both to the set A and to the line D. This contradicts the assertion that $D \cap A = \{ \langle 0, 0 \rangle \}$.

Thus, it remains to be shown that, in fact, there exists a straight line D in the plane, such that $D \cap A = \{ \langle 0, 0 \rangle \}$. In order to establish this, let us enumerate some of the properties of the set A:

1. The set A possesses at least one point that is different from the origin, $\langle 0, 0 \rangle$. This follows immediately from our assumption that L possesses a semipositive vector.

2. If $\mathbf{p} = \langle p_1, p_2 \rangle$ belongs to A, then $\lambda\mathbf{p} = \langle \lambda p_1, \lambda p_2 \rangle$ belongs to A, for every nonnegative real number λ. (The proof of this assertion is left to the reader.) Geometrically, this means that the set A is actually a collection of *rays* that emanate from the origin. We note, in passing, that such a set is called a **cone**.

3. If $\mathbf{p} = \langle p_1, p_2 \rangle$ and $\mathbf{q} = \langle q_1, q_2 \rangle$ both belong to A, then $\mathbf{p} + \mathbf{q} = \langle p_1 + q_1, p_2 + q_2 \rangle$ belongs to A. (Once again, the proof is left to the reader.) Geometrically, the properties 1–3 amount to the assertion that A is a section of the plane enclosed between two rays. Such a set is sometimes referred to as a **convex cone**.

4. If $\mathbf{p} = \langle p_1, p_2 \rangle$ belongs to A, and if $\mathbf{p} \neq \langle 0, 0 \rangle$, then $-\mathbf{p} = \langle -p_1, -p_2 \rangle$ *does not* belong to A. In other words, the set A does not contain a straight line through the origin. This assertion follows, in a straightforward manner, from the assumption that M does not possess a semipositive vector.

5. The set A is *closed*, in the sense that both of its bounding rays are contained in it. The proof of this assertion is given in an appendix at the end of the chapter.

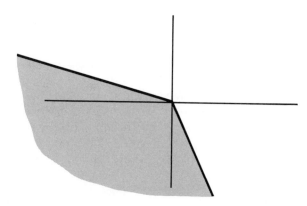

Figure 1

Properties 1–5 of the set A may be summarized as follows: The set A is a section of the plane enclosed between two rays that are both contained in A and that form an angle of less than 180° (e.g., as in Figure 1). This information leads immediately, via elementary geometry, to the conclusion that there exists in the plane a straight line whose only point in common with A is the origin. ∎

With the aid of Proposition 12.1, we shall now be able to obtain a basic result.

Proposition 12.2

Let M be a subspace of $\mathbf{R}^n$. Exactly one of the following two assertions is true:

    (i) *There exists a semipositive vector in M.*
    (ii) *There exists a positive vector in $M^\perp$.*

PROOF:

We leave it to the reader to show that (i) and (ii) cannot both be true. To see that at least one of them must be true, suppose that (i) is false. In other words, suppose that there are no semipositive vectors in M. We will show that (ii) must now be true, i.e., that $M^\perp$ must possess a positive vector. First, we consider the case where $dim\ M = n - 1$, and later on we shall turn to the case where $dim\ M < n - 1$.

Case a: If $dim\ M = n - 1$, then there exists a vector $\mathbf{t}$ in $\mathbf{R}^n$ such that M is given by

$$M = \{\mathbf{x} \,|\, \mathbf{x} \in \mathbf{R}^n \quad and \quad (\mathbf{t}, \mathbf{x}) = 0\}.$$

This follows, for example, from Proposition 7.6. We claim that the vector $\mathbf{t}$ has the following property:

$$either \quad \mathbf{t} \gg 0 \quad or \quad -\mathbf{t} \gg 0.$$

Showing this will complete the proof for Case a, because both $\mathbf{t}$ and $-\mathbf{t}$ obviously belong to $M^\perp$. Let the components of the vector $\mathbf{t}$ be denoted $t_1, \ldots, t_n$. Observe, first of all, that *none* of these components can be zero. For if $t_i = 0$, then $\mathbf{e}_i$, the i-th unit vector, belongs to M, since it satisfies $(\mathbf{t}, \mathbf{e}_i) = 0$, and this contradicts our assumption that M does not possess a semipositive vector. It remains to be shown that $t_1, \ldots, t_n$ are either all positive or all negative. Suppose not; in other words, suppose that $t_i > 0$ and $t_j < 0$ for some i and j. Construct a vector $\mathbf{x}$, with components $x_1, \ldots, x_n$, in the following manner: $x_i = -t_j$, $x_j = t_i$, and $x_k = 0$ for $k \neq i, j$. Then, $\mathbf{x}$ is semipositive and, by construction, it belongs to M. Once again, this contradicts our assumption about M.

Case b: Suppose that $dim\ M = k$, where $k < n - 1$. By Proposition 12.1, there exists a subspace V such that $V \supset M$ and $dim\ V = k + 1$, and such that V still has no semipositive vectors. Indeed, by repeatedly applying Proposition 12.1, we arrive at the conclusion that there exists a subspace, say $\hat{M}$, such that $\hat{M} \supset M$, $dim\ \hat{M} = n - 1$, and such that $\hat{M}$ has no semipositive vectors. Therefore, using Case a, we find that the subspace $\hat{M}^\perp$ must possess a positive vector. But, from the definition of the orthogonal complement, together with the assertion $\hat{M} \supset M$, it follows immediately that $\hat{M}^\perp \subset M^\perp$. Hence, $M^\perp$ must possess a positive vector. ∎

Proposition 12.2 has the following immediate application.

Proposition 12.3

Let A be a real $m \times n$ matrix. Then, exactly one of the following two assertions is true:

  (i) *The equation $Ax = 0$ has a semipositive solution;*
  (ii) *The inequality $A'y \gg 0$ has a solution.*

Similarly, of the two assertions

  (iii) *The equation $Ax = 0$ has a positive solution;*
  (iv) *The inequality $A'y > 0$ has a solution;*

exactly one is true.

PROOF:
Left to the reader. ∎

Proposition 12.3 deals, on one hand, with the existence of semipositive (or positive) solutions for homogeneous linear equations and, on the other hand, with the existence of solutions for homogeneous linear inequalities. However, from the dichotomy (iii)–(iv) in Proposition 12.3, we can get a useful result also on the existence of positive solutions for non-homogeneous linear equations. Specifically, we have:

Proposition 12.4

Let A be a real $m \times n$ matrix, and let b belong to R^m. Assume that the equation $Ax = b$ has a solution. Then, exactly one of the following assertions is true:

  (i) *The equation $Ax = b$ has a positive solution.*
  (ii) *The inequality $A'y > 0$ has a solution satisfying $(b, y) \leq 0$.*

PROOF:

Let **B** be the real $m \times (n + 1)$ matrix which is obtained by adjoining the vector $-\mathbf{b}$ to the matrix **A**: $\mathbf{B} = [\mathbf{A}, -\mathbf{b}]$. Applying Proposition 12.3 to the matrix **B**, we find that either the equation $\mathbf{Bw} = \mathbf{0}$ has a positive solution or the inequality $\mathbf{B'y} > \mathbf{0}$ has a solution. If the equation $\mathbf{Bw} = \mathbf{0}$ has a positive solution, then it clearly has a solution **w** for which $w_{n+1} = 1$. Therefore, we conclude that *either* the equation $\mathbf{Ax} = \mathbf{b}$ has a positive solution *or* the inequalities $\mathbf{A'y} \geqq \mathbf{0}$ and $(-\mathbf{b}, \mathbf{y}) \geqq 0$ have a solution with the symbol $>$ appearing in at least one of them. But, by Proposition 11.2, if there exists a vector **y** such that $\mathbf{A'y} = \mathbf{0}$ and $(-\mathbf{b}, \mathbf{y}) > 0$, then the equation $\mathbf{Ax} = \mathbf{b}$ has no solution at all. This leads immediately to the desired result. ∎

In order to derive a condition for the existence of nonnegative (rather than positive) solutions for non-homogeneous linear equations, we must turn, once again, to a pair of preliminary propositions. These propositions are very similar, both in content and in proof, to Propositions 12.1 and 12.2.

Let **x** be a member of the space $\mathbf{R}^n$, and let its components be denoted $x_1, \ldots, x_n$. Consider an arbitrary integer, j, satisfying $1 \leqq j \leqq n$. We shall say that the vector **x** is *j-positive* iff $\mathbf{x} \geqq \mathbf{0}$ and $x_j > 0$. If **x** is j-positive, then we shall write $\mathbf{x} \underset{j}{>} \mathbf{0}$.

Proposition 12.5.

Let M be a k-dimensional subspace of $\mathbf{R}^n$, and let j be an integer satisfying $1 \leqq j \leqq n$. Assume that there are no j-positive vectors in M. If $k < n - 1$, then M can be embedded in another subspace, say V, such that $\dim V = k + 1$ and such that V still has no j-positive vectors.

PROOF:

Let us proceed along the lines of the proof of Proposition 12.1. Let L be a subspace containing M, with $\dim L = k + 2$, and let $B = \langle \mathbf{x}_1, \ldots, \mathbf{x}_{k+2} \rangle$ be a basis for L, whose first k components constitute a basis for M. Define a set, A, by

$$A = \left\{ \langle \alpha_{k+1}, \alpha_{k+2} \rangle \middle| \begin{array}{l} \textit{There exists a k-tuple } \langle \alpha_1, \ldots, \alpha_k \rangle \\ \textit{for which } \sum_{i=1}^{k+2} \alpha_i \mathbf{x}_i \underset{j}{>} \mathbf{0} \end{array} \right\}.$$

In this case, the origin, $\langle 0, 0 \rangle$, does not belong to A. Our proposition will be proved if we can find a straight line in the plane that passes through the origin and that does not intersect the set A. (We refer the reader to the proof of Proposition 12.1, to see why this is so.) In order to establish the existence of such a straight line, we must, once again, study the properties of the set A. In all cases, the proof that A does, in fact, have the stated property is quite straightforward, and it will therefore be left to the reader.

1. The set A is nonempty. (This is true because we may assume, without loss of generality, that L does possess a j-positive vector.)
2. If $\mathbf{p} \in A$, then $\lambda\mathbf{p} \in A$, for every real number λ satisfying $\lambda > 0$.
3. If $\mathbf{p} \in A$ and $\mathbf{q} \in A$, then $\mathbf{p} + \mathbf{q} \in A$.
4. If $\mathbf{p} \in A$, then $-\mathbf{p} \notin A$.

So far, except for the fact that $\langle 0, 0 \rangle \notin A$, we find that the set A has exactly the same properties that the set A in Proposition 12.1 had. In other words, A is a section of the plane enclosed between two rays that emanate from the origin and, furthermore, the set A does not contain a straight line through the origin. However, in Proposition 12.1, the set A had a fifth property—it contained both of its bounding rays. This property does not hold in the present case. Therefore, we must try and manage without it. Recall that what we are trying to show is that there exists a straight line through the origin with no points in common with A. Now, the *only* case in which A satisfies the properties 1–4 and in which such a straight line fails to exist is the case where A is enclosed between two rays that form an angle of 180° and where *one* of the bounding rays is contained in A, but not the other (as in Figure 2). Thus, we must show that the situation depicted in Figure 2 does not arise.

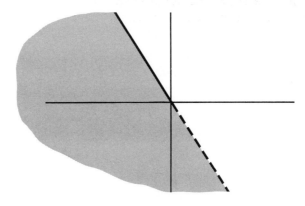

Figure 2

Let $\mathbf{p} = \langle p_1, p_2 \rangle$ be a point on the bounding ray that is *not* contained in A. Since $\mathbf{p} \notin A$, there does not exist a k-tuple $\langle \alpha_1, \ldots, \alpha_k \rangle$ such that

$$\sum_{i=1}^{k} \alpha_i \mathbf{x}_i + p_1 \mathbf{x}_{k+1} + p_2 \mathbf{x}_{k+2} \underset{j}{>} \mathbf{0}.$$

However, it can be shown that there *does* exist a k-tuple $\langle \alpha_1, \ldots, \alpha_k \rangle$ such that

(*) $$\sum_{i=1}^{k} \alpha_i \mathbf{x}_i + p_1 \mathbf{x}_{k+1} + p_2 \mathbf{x}_{k+2} \geqq \mathbf{0}.$$

This follows from the fact that the point $\mathbf{p}$, while it does not belong to the set A, is the limit of points that do belong to A. Now, the point $-\mathbf{p}$ $(= \langle -p_1, -p_2 \rangle)$ is located on the opposite ray, and this ray, by assumption, *is* contained in A. Therefore, there exists a k-tuple $\langle \beta_1, \ldots, \beta_k \rangle$ such that

(**) $$\sum_{i=1}^{k} \beta_i \mathbf{x}_i - p_1 \mathbf{x}_{k+1} - p_2 \mathbf{x}_{k+2} \underset{j}{>} \mathbf{0}.$$

But now, if we add the equations (*) and (**) together, we find that

$$\sum_{i=1}^{k} (\alpha_i + \beta_i) \mathbf{x}_i \underset{j}{>} \mathbf{0},$$

which contradicts the hypothesis that M does not possess a j-positive vector. Thus, the situation depicted in Figure 2 cannot arise. ∎

Proposition 12.6

Let M be a subspace of $\mathbf{R}^n$, and let j be an integer satisfying $1 \leqq j \leqq n$. Then, exactly one of the following assertions is true:

    (i) *There exists a j-positive vector in M.*
    (ii) *There exists a j-positive vector in $M^{\perp}$.*

PROOF:

It is immediate that (i) and (ii) cannot both be true. To see that at least one of them is true, assume that (i) is false, i.e., that there are no j-positive vectors in M.

Case a: Suppose that $\dim M = n - 1$. Then, there exists in $\mathbf{R}^n$ a vector, say $\mathbf{t}$, such that

$$M = \{\mathbf{x} \mid \mathbf{x} \in \mathbf{R}^n \quad and \quad (\mathbf{t}, \mathbf{x}) = 0\}.$$

Since M does not possess a j-positive vector, it certainly does not possess a positive vector. Therefore, by Proposition 12.2, $M^\perp$ must possess a semipositive vector. And since all members of $M^\perp$ are multiples of the vector $\mathbf{t}$, we conclude that $\mathbf{t}$ can be chosen to be semipositive. It now follows that the assertion (ii) is true iff the j-th component of $\mathbf{t}$ is positive. But if the j-th component of $\mathbf{t}$ is zero, then $\mathbf{e}_j$, the j-th unit vector, belongs to M, since it satisfies $(\mathbf{t}, \mathbf{e}_j) = 0$. This contradicts our assumption that (i) is false.

Case b: *dim* $M < n - 1$. This case can be reduced to Case *a*, by repeated application of Proposition 12.5. ∎

We are now ready to state the final proposition of this chapter. This proposition is known as the *lemma of Farkas*.

Proposition 12.7

Let $\mathbf{A}$ *be a real* $m \times n$ *matrix, and let* $\mathbf{b}$ *belong to* $\mathbf{R}^m$. *Then, exactly one of the following assertions is true:*

(i) *The equation* $\mathbf{A}\mathbf{x} = \mathbf{b}$ *has a nonnegative solution.*

(ii) *The inequality* $\mathbf{A}'\mathbf{y} \geqq \mathbf{0}$ *has a solution satisfying the condition* $(\mathbf{y}, \mathbf{b}) < 0$.

PROOF:

Consider the matrix $\mathbf{B} = [\mathbf{A}, -\mathbf{b}]$, which is obtained by adjoining the vector $-\mathbf{b}$ to the matrix $\mathbf{A}$. Let M be the null space of $\mathbf{B}$, and apply Proposition 12.6, with $j = n + 1$. ∎

Propositions 12.3, 12.4, and 12.7 are often referred to as theorems of *alternatives* or as theorems of *duality*. They have applications in linear programming, in game theory, and in other theoretical branches of the social sciences.

APPENDIX

SUPPLEMENT TO THE PROOF OF PROPOSITION 12.1

Consider a sequence, $\langle \alpha_{l+1}^h, \alpha_{k+2}^h \rangle$ $h = 1, 2, \ldots$, of non-null points of A. Suppose that this sequence converges, as $h \to \infty$, to a point $\langle \alpha_{k+1}^*, \alpha_{k+2}^* \rangle$ $\neq \langle 0, 0 \rangle$. For each h, there exists a k-tuple, $\langle \alpha_1^h, \ldots, \alpha_k^h \rangle$, such that

$$\sum_{i=1}^{k+2} \alpha_i^h \mathbf{x}_i \geqq \mathbf{0}.$$

For every $(k + 2)$-tuple $\langle \alpha_1^h, \ldots, \alpha_{k+2}^h \rangle$, consider the quantity $\sqrt{\sum (\alpha_i^h)^2}$. This quantity cannot be zero since, by assumption, the pairs $\langle \alpha_{k+1}^h, \alpha_{k+2}^h \rangle$ are non-null. Therefore, we may divide every component of the $(k + 2)$-tuple $\langle \alpha_1^h, \ldots, \alpha_{k+2}^h \rangle$ by $\sqrt{\sum (\alpha_i^h)^2}$, and we may do so for every h. The result is a new, normalized, sequence of $(k + 2)$-tuples, $\langle \beta_1^h, \ldots, \beta_{k+2}^h \rangle$ for $h = 1$, $2, \ldots$, with the property that

$$\sum (\beta_i^h)^2 = 1 \quad \text{for all} \quad h.$$

It follows from this last equation that there exists a subsequence of the sequence of $(k + 2)$-tuples $\langle \beta_1^h, \ldots, \beta_{k+2}^h \rangle$ which converges to a $(k + 2)$-tuple $\langle \beta_1^*, \ldots, \beta_{k+2}^* \rangle$. For convenience, let us say that the sequence of $(k + 2)$-tuples $\langle \beta_1^h, \ldots, \beta_{k+2}^h \rangle$ itself converges to $\langle \beta_1^*, \ldots, \beta_{k+2}^* \rangle$. Thus, for each h, we have $\sum \beta_i^h x_i \geqq 0$, and since the limit of a sequence of nonnegative real numbers is nonnegative, we get $\sum \beta_i^* x_i \geqq 0$. This means that $\langle \beta_{k+1}^*, \beta_{k+2}^* \rangle \in A$. It is clear also that $\langle \beta_{k+1}^*, \beta_{k+2}^* \rangle$ must lie on the same ray as $\langle \alpha_{k+1}^*, \alpha_{k+2}^* \rangle$. In other words, $\langle \beta_{k+1}^*, \beta_{k+2}^* \rangle$ must be a nonnegative multiple of $\langle \alpha_{k+1}^*, \alpha_{k+2}^* \rangle$. We claim that $\langle \beta_{k+1}^*, \beta_{k+2}^* \rangle$ is, in fact, a *positive* multiple of $\langle \alpha_{k+1}^*, \alpha_{k+2}^* \rangle$, i.e., that $\langle \beta_{k+1}^*, \beta_{k+2}^* \rangle \neq \langle 0, 0 \rangle$. Suppose for a moment, contrary to this claim, that $\langle \beta_{k+1}^*, \beta_{k+2}^* \rangle = \langle 0, 0 \rangle$. Then, it follows that

$$\sum_{i=1}^{k} \beta_i^* x_i \geqq 0.$$

And since $\sum (\beta_i^*)^2 = 1$, we may actually write

$$\sum_{i=1}^{k} \beta_i^* x_i > 0.$$

But this contradicts the hypothesis that the subspace M does not possess a semipositive vector. Thus, $\langle \beta_{k+1}^*, \beta_{k+2}^* \rangle \neq \langle 0, 0 \rangle$. It now follows that $\langle \alpha_{k+1}^*, \alpha_{k+2}^* \rangle$ is a positive multiple of $\langle \beta_{k+1}^*, \beta_{k+2}^* \rangle$, so $\langle \alpha_{k+1}^*, \alpha_{k+2}^* \rangle$ must belong to the set A. This means that A contains both of its bounding rays.

PROBLEMS

12.1. In the proof of Proposition 12.1, show that the set A satisfies the properties 1–4.

12.2. Show that, in Proposition 12.2, the assertions (i) and (ii) cannot both be true.

12.3. Prove Proposition 12.3.

12.4. Let $\mathbf{A}$ be a real $m \times n$ matrix. Show that exactly one of the following two assertions is true:

     (i) The equation $\mathbf{Ax} = \mathbf{0}$ has a solution with a positive first component.

    (ii) There exist a real number, λ, and a vector, $\mathbf{y}$ in $\mathbf{R}^m$, such that $\lambda > 0$ and $\mathbf{A'y} = \mathbf{e}_1$, where $\mathbf{e}_1$ is the first unit vector in $\mathbf{R}^n$.

Note that, in (i), $\mathbf{x}$ is required to have a positive first component, but the other components are entirely unrestricted.

13 Permutations

A permutation is a one-to-one function on a finite set of distinct objects back onto the same finite set. Since the nature of the objects that comprise this set is of no relevance, we usually speak about a permutation as a function on the set of the first n positive integers. Formally, we define a **permutation of the integers 1, 2, ..., n** as a one-to-one function on the set $\{1, 2, \ldots, n\}$ onto itself.

The reader may be familiar with the use of n-tuple notation in the description of permutations. For example, one often sees an object like $\langle 5, 1, 3, 2, 4\rangle$ referred to as a permutation of the integers 1, 2, 3, 4, 5. Strictly speaking, however, the phrase "the permutation $\langle 5, 1, 3, 2, 4\rangle$" is merely a convenient abbreviation for the more cumbersome phrase "the permutation p of the integers 1, 2, 3, 4, 5, such that $p(1) = 5$, $p(2) = 1$, $p(3) = 3$, $p(4) = 2$, and $p(5) = 4$." In the present chapter, we shall study permutations only insofar as they will enter into our forthcoming discussion of determinants. In this context, it turns out that the more cumbersome notation is also more convenient, so we shall, for the most part, refrain from using n-tuples to denote permutations.

The propositions on permutations that we shall need are all fairly simple propositions, and for this reason we shall omit most of the proofs. However, the reader may wish to work out the proofs for himself, as this provides a useful exercise in the manipulation of simple functions. In some cases, the text will contain a rough sketch of how a proof might be arrived at.

Proposition 13.1

The following assertions are true:

    (i) *The number of permutations of the integers* $1, 2, \ldots, n$ *is finite. Specifically there are exactly* $n!$ *(read: n factorial) distinct permutations of the integers* $1, 2, \ldots, n$, *where* $n!$ *is defined by* $n! = 1 \cdot 2 \cdot 3 \cdots n$.

    (ii) *The identity function on* $\{1, 2, \ldots, n\}$ *is a permutation of the integers* $1, 2, \ldots, n$.

    (iii) *If p is a permutation of the integers* $1, 2, \ldots, n$, *then the inverse function,* p^{-1} *(exists and) is a permutation of the integers* $1, 2, \ldots, n$.

    (iv) *If p and p' are permutations of the integers* $1, 2, \ldots, n$, *then the composition* $p \circ p'$ *(exists and) is a permutation of the integers* $1, 2, \ldots, n$.

PROOF:

All four assertions are immediate. ∎

119

We now proceed to isolate, within the set of all permutations of the integers $1, 2, \ldots, n$, a subset consisting of certain very simple permutations. These special permutations will assume a role similar to that of elementary matrices in Chapter 11.

Let k be a member of the set $\{1, 2, \ldots, n\}$ and assume that $k \neq n$. If π is a permutation of the integers $1, 2, \ldots, n$ such that

$$\pi(k) = k + 1$$
$$\pi(k + 1) = k$$
$$\pi(i) = i \quad \text{if} \quad i \neq k \quad \text{and} \quad i \neq k + 1,$$

then π is referred to as an **inversion**. We say that π **inverts** k and $k + 1$.

Proposition 13.2

Every permutation of the integers $1, 2, \ldots, n$ is the composition of a finite number of inversions.

PROOF:
Left to the reader. ∎

Proposition 13.3

Let I be the identity function on the set $\{1, 2, \ldots, n\}$. If $\pi_1, \pi_2, \ldots, \pi_k$ are inversions such that

$$\pi_1 \circ \pi_2 \circ \cdots \circ \pi_k = I,$$

then k is even.

SKETCH OF PROOF:
Define the permutations $p_1, p_2, \ldots, p_k$ as follows:

$$p_i = p_{i-1} \circ \pi_i \quad \text{for} \quad i = 2, 3, \ldots, k$$
$$p_1 = \pi_1.$$

Now look at the sequence of integers $p_1^{-1}(n), p_2^{-1}(n), \ldots, p_k^{-1}(n)$. There are three possibilities:

(i) $p_i^{-1}(n) = p_{i-1}^{-1}(n);$

(ii) $p_i^{-1}(n) = p_{i-1}^{-1}(n) - 1;$

(iii) $p_i^{-1}(n) = p_{i-1}^{-1}(n) + 1.$

In the first case, we say that the inversion π_i leaves n unaffected. In the second case, we say that π_i shifts n to the left. Finally, in the third case, we say that π_i shifts n to the right. Now, by hypothesis, $p_k = I$ and therefore $p_k^{-1}(n) = n$. From this it follows that there are exactly as many inversions in the sequence $\pi_1, \pi_2, \ldots, \pi_k$ that shift n to the left as there are inversions in this sequence that shift n to the right. Thus, the total number of inversions that shift n (whether to the left or to the right) must be even. Now we proceed to remove from the sequence $\pi_1, \pi_2, \ldots, \pi_k$ those inversions that shift n (to the left or to the right) and, correspondingly, we remove the element n from the set $\{1, 2, \ldots, n\}$. It is easy to see that the remaining inversions may now be looked upon as permutations of the integers $1, 2, \ldots, n - 1$ and, indeed, that they form a sequence of inversions whose composition is equal to the identity function on $\{1, 2, \ldots, n - 1\}$. The next step is to isolate, among these remaining inversions, those particular ones that shift the integer $n - 1$ to the left or to the right, and to notice that they too must be even in number. Continuing in this fashion, we get the desired result. ∎

Our next proposition is essentially a corollary of the foregoing.

Proposition 13.4

If a permutation of the integers $1, 2, \ldots, n$ can be written as the composition of an even number of inversions, then it cannot be written as the composition of an odd number of inversions. Similarly, if it can be written as the composition of an odd number of inversions, then it cannot be written as the composition of an even number of inversions.

PROOF:

Left to the reader. ∎

A permutation of the integers $1, 2, \ldots, n$ is said to be **even** if it can be expressed as the composition of an even number of inversions, and it is said to be **odd** if it can be expressed as the composition of an odd number of inversions. In view of Proposition 13.2, every permutation of the integers $1, 2, \ldots, n$ is either even or odd, and, in view of Proposition 13.4, no permutation of the integers $1, 2, \ldots, n$ is both even and odd. Every inversion is, obviously, an odd permutation and, by Proposition 13.3, the identity is an even permutation.

If p is a permutation of the integers $1, 2, \ldots, n$, then we define the **sign** of p, written *sgn p*, as follows:

$$sgn\, p = 1 \quad \text{if } p \text{ is even;}$$
$$= -1 \text{ if } p \text{ is odd.}$$

Proposition 13.5

Let p and p' be two permutations of the integers 1, 2, . . . , n. Then,

(i) $sgn\,(p \circ p') = (sgn\,p)(sgn\,p')$;
(ii) $sgn\,p^{-1} = sgn\,p$.

PROOF:

Left to the reader. ∎

Let j and k be two distinct members of the set $\{1, 2, \ldots, n\}$, and let p be the permutation of the integers $1, 2, \ldots, n$ defined by

$$p(k) = j$$
$$p(j) = k$$
$$p(i) = i \quad \text{if} \quad i \neq j \quad \text{and} \quad i \neq k.$$

Then, p is referred to as a **transposition**. We say that p **transposes** j and k.

Proposition 13.6

Every transposition is an odd permutation.

PROOF:

Left to the reader. ∎

In our forthcoming discussion of determinants, we shall have occasion to consider the question of converting a permutation of the integers 1, 2, . . . , n into a permutation of the integers $1, 2, \ldots, n - 1$. Specifically, if p is a permutation of the integers $1, 2, \ldots, n$, and if $p(j) = k$ for some j and k, then we shall be interested in deleting the integer j from the domain of p, deleting the integer k from the range of p, and using what remains to construct a permutation of the integers $1, 2, \ldots, n - 1$. The simplest case is the following: Let p be a permutation of the integers $1, 2, \ldots, n$, and assume that $p(n) = n$. If we delete the integer n both from the domain of p and from its range, we arrive, without further ado, at a permutation of the integers $1, 2, \ldots, n - 1$. We shall denote this new permutation p^{nn}, to signify that the integer n is being deleted both from the domain and from the range of p. Note that $sgn\,p^{nn} = sgn\,p$. Next in complexity is the case of a permutation p of the integers $1, 2, \ldots, n$, where $p(j) = n$, and where we are interested in deleting the integer j from the domain of p and the integer n from its range. But deleting j and n is not enough in this case, because what remains after this deletion is a function on the set $\{1, 2, \ldots, j - 1, j + 1, \ldots, n\}$ onto the set $\{1, 2, \ldots, n - 1\}$ and, unless $j = n$, this function is not a permutation of the integers $1, 2, \ldots, n - 1$. What needs to be done with this function, to obtain the desired permutation, is to "close the ranks" of the set $\{1, 2, \ldots, j - 1, j + 1, \ldots, n\}$. This task is achieved by first composing the permutation p with a sequence of inversions, and then deleting the integer n from the domain and from the range of this composition. For $k = 1, 2, \ldots, n - 1$, let π_k be the inversion defined by

$$\pi_k(k) = k + 1$$
$$\pi_k(k + 1) = k$$
$$\pi_k(i) = i \quad \text{if} \quad i \neq k \quad \text{and} \quad i \neq k + 1.$$

Now, define the permutation $\hat{p}$ as follows:

$$\hat{p} = p \circ (\pi_j \circ \pi_{j+1} \circ \cdots \circ \pi_{n-1}).$$

It follows immediately from our assumptions on p that $\hat{p}(n) = n$. Therefore, we are in a position to delete the integer n from the domain of $\hat{p}$ and from its range, thus obtaining the desired permutation, to be denoted p^{jn}, of the integers $1, 2, \ldots, n - 1$.

An example may be in order here. Let p be the permutation of the integers $1, 2, 3, 4, 5$ such that

$$p(1) = 4, p(2) = 5, p(3) = 1, p(4) = 3, p(5) = 2.$$

From the domain of p we now delete the integer 2 and from its range we delete the integer 5. We get the following function, call it f, on the set $\{1, 3, 4, 5\}$ onto the set $\{1, 2, 3, 4\}$

$$f(1) = 4, f(3) = 1, f(4) = 3, f(5) = 2.$$

Closing the ranks of the domain of f means converting 3 into 2, 4 into 3, and 5 into 4. The result is the permutation, p^{25}, defined by

$$p^{25}(1) = 4, p^{25}(2) = 1, p^{25}(3) = 3, p^{25}(4) = 2.$$

The reader can easily verify that this permutation is, indeed, equal to $p \circ (\pi_2 \circ \pi_3 \circ \pi_4)$.

Let us now proceed to the general case. A permutation p of the integers $1, 2, \ldots, n$ is given, such that $p(j) = k$ for some j and k belonging to the set $\{1, 2, \ldots, n\}$. We are interested in the permutation of the integers $1, 2, \ldots, n - 1$ which emerges after one deletes j from the domain of p and k from its range, and after the ranks of both the domain and the range are closed. To obtain this permutation, we must first compose p with *two* sequences of inversions, one on the right (as before) and one on the left. Specifically, let $\hat{p}$ be defined by the formula

$$\hat{p} = (\pi_{n-1} \circ \cdots \circ \pi_{k+1} \circ \pi_k) \circ p \circ (\pi_j \circ \pi_{j+1} \circ \cdots \circ \pi_{n-1}).$$

Then, $\hat{p}(n) = n$ and we can obtain the desired permutation by deleting the integer n from the domain of $\hat{p}$ as well as from its range. We denote this permutation p^{jk}.

Proposition 13.7

If p is a permutation of the integers 1, 2, . . . , n, and if p^{jk} is the permutation that is derived from p in the manner just described, then

$$sgn \, p^{jk} = (-1)^{j+k} \, sgn \, p.$$

Proof:

Left to the reader. ∎

To end this chapter, let us describe a simple procedure by which, given the permutation p, one can find the permutation p^{jk}.

Let p be an arbitrary permutation of the integers 1, 2, . . . , n, and define a real $n \times n$ matrix $\mathbf{P}$ as follows:

$$\mathbf{P} = \langle \mathbf{e}_{p(1)}, \mathbf{e}_{p(2)}, \ldots, \mathbf{e}_{p(n)} \rangle.$$

$\mathbf{P}$ is the matrix whose columns are the unit vectors, ordered according to the permutation p. We refer to $\mathbf{P}$ as the matrix of the permutation p.† For example, if p is the permutation of the integers 1, 2, 3, 4, 5 such that

$$p(1) = 4, p(2) = 5, p(3) = 1, p(4) = 3, p(5) = 2,$$

then $\mathbf{P}$ is given by

$$\mathbf{P} = \begin{bmatrix} 0 & 0 & 1 & 0 & 0 \\ 0 & 0 & 0 & 0 & 1 \\ 0 & 0 & 0 & 1 & 0 \\ 1 & 0 & 0 & 0 & 0 \\ 0 & 1 & 0 & 0 & 0 \end{bmatrix}.$$

Now suppose that, for this specific example, we wish to find the permutation

---

† Some people define the matrix of a permutation p as the matrix whose *rows* (rather than columns) are the unit vectors, ordered according to p.

p^{43}. To do this, we simply cross out the fourth column and the third row of **P**, thus obtaining the matrix

$$\begin{bmatrix} 0 & 0 & 1 & 0 \\ 0 & 0 & 0 & 1 \\ 1 & 0 & 0 & 0 \\ 0 & 1 & 0 & 0 \end{bmatrix}.$$

This last matrix is clearly the matrix of some permutation of the integers 1, 2, 3, 4. It is not difficult to prove that it is, in fact, the matrix of the permutation p^{43}. In other words, p^{43} is given by

$$p^{43}(1) = 3, \, p^{43}(2) = 4, \, p^{43}(3) = 1, \, p^{43}(4) = 2.$$

In the general case, if we have a permutation p of the integers $1, 2, \ldots, n$, and if **P** is the matrix of p, then, by deleting the j-th column and the k-th row of **P**, we obtain a real $(n - 1) \times (n - 1)$ matrix, call it $\mathbf{P}_{jk}$, and this is precisely the matrix of the permutation p^{jk}. Let us sketch the proof of this assertion. Recall that p^{jk} is obtained by deleting the integer n from the domain and the range of a permutation $\hat{p}$, defined by

$$\hat{p} = (\pi_{n-1} \circ \ldots \circ \pi_k) \circ p \circ (\pi_j \circ \ldots \circ \pi_{n-1}).$$

Let $\mathbf{\Pi}_i$ be the matrix of the inversion π_i, and define a matrix $\hat{\mathbf{P}}$ by

$$\hat{\mathbf{P}} = \mathbf{\Pi}_{n-1}\mathbf{\Pi}_{n-2} \cdots \mathbf{\Pi}_k \mathbf{P} \mathbf{\Pi}_j \mathbf{\Pi}_{j+1} \cdots \mathbf{\Pi}_{n-1}.$$

Then, it is easy to show that $\hat{\mathbf{P}}$ is the matrix of the permutation $\hat{p}$ and that, furthermore, when the n-th column and the n-th row of $\hat{\mathbf{P}}$ are deleted, one is left precisely with the matrix $\mathbf{P}_{jk}$. This means that $\mathbf{P}_{jk}$ is the matrix of the permutation p^{jk}.

PROBLEM

Prove all the propositions of this chapter in detail.

14 Determinants

Let f be a function that carries n-tuples of the form $\langle \mathbf{x}_1, \mathbf{x}_2, \ldots, \mathbf{x}_n \rangle$, where the $\mathbf{x}_i$'s are members of the space $\mathbf{R}^n$, into real numbers. In other words, f is a function on the n-fold Cartesian product $\mathbf{R}^n \times \cdots \times \mathbf{R}^n$ to $\mathbf{R}$. We say that f is an **n-linear form** in $\mathbf{R}^n$ iff, for every member j of the set $\{1, 2, \ldots, n\}$ and for any fixed vectors

$$\mathbf{x}_1^0, \mathbf{x}_2^0, \ldots, \mathbf{x}_{j-1}^0, \mathbf{x}_{j+1}^0, \ldots, \mathbf{x}_n^0,$$

all in $\mathbf{R}^n$, the following two conditions hold:

(i) $f(\mathbf{x}_1^0, \ldots, \mathbf{x}_{j-1}^0, \lambda \mathbf{x}, \mathbf{x}_{j+1}^0, \ldots, \mathbf{x}_n^0)$
$= \lambda f(\mathbf{x}_1^0, \ldots, \mathbf{x}_{j-1}^0, \mathbf{x}, \mathbf{x}_{j+1}^0, \ldots, \mathbf{x}_n^0);$

(ii) $f(\mathbf{x}_1^0, \ldots, \mathbf{x}_{j-1}^0, \mathbf{x} + \mathbf{y}, \mathbf{x}_{j+1}^0, \ldots, \mathbf{x}_n^0)$
$= f(\mathbf{x}_1^0, \ldots, \mathbf{x}_{j-1}^0, \mathbf{x}, \mathbf{x}_{j+1}^0, \ldots, \mathbf{x}_n^0)$
$+ f(\mathbf{x}_1^0, \ldots, \mathbf{x}_{j-1}^0, \mathbf{y}, \mathbf{x}_{j+1}^0, \ldots, \mathbf{x}_n^0)$

where $\mathbf{x}$ and $\mathbf{y}$ are any members of $\mathbf{R}^n$ and where λ is any real number. In other words, f is an n-linear form in $\mathbf{R}^n$ iff it is, so to speak, a linear transformation in each variable separately. (This property is often referred to as **multilinearity**.)

An n-tuple of the form $\langle x_1, x_2, \ldots, x_n \rangle$, with components in $\mathbf{R}^n$, is, of course, the same thing as a real $n \times n$ matrix. Whether we should identify $\langle x_1, x_2, \ldots, x_n \rangle$ with the matrix whose *columns* are the vectors $x_1, x_2, \ldots,$ x_n or with the matrix whose *rows* are the vectors $x_1, x_2, \ldots, x_n$ is entirely a matter of convention. In the present volume, our practice has been, and so it will remain, to identify the n-tuple $\langle x_1, x_2, \ldots, x_n \rangle$ with the matrix whose *columns* are the vectors $x_1, x_2, \ldots, x_n$. Accordingly, a function on the set of all real $n \times n$ matrices to the real number system is an n-linear form in $\mathbf{R}^n$ iff it is a linear transformation "in each column separately." Example: Let X be an arbitrary real $n \times n$ matrix, with entries x_{ij}, and let ξ_j be the sum of the entries in the j-th column of X. In other words, $\xi_j = \sum_{i=1}^{n} x_{ij}$ for $j = 1, 2, \ldots, n$. Then, the function f defined by

$$f(X) = \xi_1 \xi_2 \cdots \xi_n$$

is an n-linear form in $\mathbf{R}^n$.

An n-linear form f in $\mathbf{R}^n$ is said to be **alternating** iff

$$f(x_{p(1)}, \ldots, x_{p(n)}) = (sgn\ p) f(x_1, \ldots, x_n)$$

for every permutation p of the integers $1, 2, \ldots, n$. We shall now proceed to develop the determinant function as an example of an alternating n-linear form in $\mathbf{R}^n$. Moreover, we shall see that, apart from a constant of multiplication, the determinant function is the *only* example of an alternating n-linear form in $\mathbf{R}^n$. But first, we need the following propositions.

Proposition 14.1

Let f be an alternating n-linear form in $\mathbf{R}^n$, and let $\mathbf{X} = \langle \mathbf{x}_1, \mathbf{x}_2, \ldots, \mathbf{x}_n \rangle$ be a real $n \times n$ matrix. If there exist two distinct members, j and k, of the set $\{1, 2, \ldots, n\}$ such that $\mathbf{x}_j = \mathbf{x}_k$, that is, if $\mathbf{X}$ has two identical columns, then $f(\mathbf{X}) = 0$.

PROOF:
By the fact that $\mathbf{x}_j = \mathbf{x}_k$, we have

$$f(\mathbf{x}_1, \ldots, \mathbf{x}_j, \ldots, \mathbf{x}_k, \ldots, \mathbf{x}_n) = f(\mathbf{x}_1, \ldots, \mathbf{x}_k, \ldots, \mathbf{x}_j, \ldots, \mathbf{x}_n).$$

On the other hand, the fact that f is alternating, together with Proposition 13.6, implies that

$$f(\mathbf{x}_1, \ldots, \mathbf{x}_j, \ldots, \mathbf{x}_k, \ldots, \mathbf{x}_n) = -f(\mathbf{x}_1, \ldots, \mathbf{x}_k, \ldots, \mathbf{x}_j, \ldots, \mathbf{x}_n).$$

Thus, $f(\mathbf{X}) = -f(\mathbf{X})$, which can only be true if $f(\mathbf{X}) = 0$. ∎

Corollary

Let f be an alternating n-linear form in $\mathbf{R}^n$, and let $\mathbf{X} = \langle \mathbf{x}_1, \mathbf{x}_2, \ldots, \mathbf{x}_n \rangle$ be a real $n \times n$ matrix. If $\mathbf{X}$ is singular, then $f(\mathbf{X}) = 0$.

PROOF:
We know that $\mathbf{X}$ is singular iff one of its columns is a linear combination of the others. For simplicity, let us assume that $\mathbf{x}_1$ is a linear combination of $\mathbf{x}_2, \mathbf{x}_3, \ldots, \mathbf{x}_n$:

$$\mathbf{x}_1 = \sum_{j=2}^{n} \alpha_j \mathbf{x}_j$$

for some real numbers $\alpha_2, \alpha_3, \ldots, \alpha_n$. Using the fact that f is an n-linear form, we obtain

$$
\begin{aligned}
f(\mathbf{X}) &= f(\mathbf{x}_1, \mathbf{x}_2, \ldots, \mathbf{x}_n) \\
&= f(\textstyle\sum \alpha_j \mathbf{x}_j, \mathbf{x}_2, \mathbf{x}_3, \ldots, \mathbf{x}_n) \\
&= \alpha_2 f(\mathbf{x}_2, \mathbf{x}_2, \mathbf{x}_3, \ldots, \mathbf{x}_n) \\
&\quad + \alpha_3 f(\mathbf{x}_3, \mathbf{x}_2, \mathbf{x}_3, \ldots, \mathbf{x}_n) \\
&\quad + \cdots\cdots\cdots\cdots \\
&\quad + \alpha_n f(\mathbf{x}_n, \mathbf{x}_2, \mathbf{x}_3, \ldots, \mathbf{x}_n).
\end{aligned}
$$

Now, each of the members of this last sum is zero, by the foregoing proposition. ∎

Once again, let $X = \langle x_1, x_2, \ldots, x_n \rangle$ be an arbitrary real $n \times n$ matrix, and assume that the ij-th entry of X is $x_{i,j}$ (or, equivalently, that the i-th component of x_j is $x_{i,j}$). Note that, for typographical reasons, we are now writing $x_{i,j}$ with a comma separating i and j. Let f be an alternating n-linear form in $\mathbf{R}^n$. We wish to use the properties of f to find a specific expression for $f(X)$. To do this, we select a basis for $\mathbf{R}^n$ and proceed to express the vectors $x_1, x_2, \ldots, x_n$ in terms of this basis. Naturally, the easiest thing to do is to pick the *unit basis* $\langle e_1, e_2, \ldots, e_n \rangle$. Recall that

$$x_j = \sum_{i=1}^{n} x_{i,j} e_i \quad \text{for} \quad j = 1, 2, \ldots, n.$$

Using this fact for x_1, we get

$$\begin{aligned}
f(X) &= f(x_1, x_2, \ldots, x_n) \\
&= f\left(\sum_{i=1}^{n} x_{i,1} e_i, x_2, x_3, \ldots, x_n \right) \\
&= \sum_{i=1}^{n} x_{i,1} f(e_i, x_2, x_3, \ldots, x_n),
\end{aligned}$$

where we have used the fact that f is n-linear. Now let us replace the index of summation, i, by the symbol $p(1)$. The reason for this change in notation will become apparent shortly. Thus, we have

$$f(X) = \sum_{p(1)=1}^{n} x_{p(1),1} f(e_{p(1)}, x_2, \ldots, x_n).$$

We emphasize again that $p(1)$ is, for the time being, just a dummy index whose function is precisely the same as that of the symbol i in the earlier formula. Proceeding now to x_2 and introducing the dummy index $p(2)$, we get

$$f(X) = \sum_{p(1)=1}^{n} x_{p(1),1} \sum_{p(2)=1}^{n} x_{p(2),2} f(e_{p(1)}, e_{p(2)}, x_3, \ldots, x_n).$$

If we continue in this fashion until all the x_j's have been expressed in terms of the unit vectors, we finally get

$$f(X) = \sum_{p(1)=1}^{n} \sum_{p(2)=1}^{n} \cdots \sum_{p(n)=1}^{n} x_{p(1),1} x_{p(2),2} \cdots x_{p(n),n} f(e_{p(1)}, e_{p(2)}, \ldots, e_{p(n)}).$$

This last equation involves multiple summation, and the reader may require a few minutes to become familiar with it. Perhaps the best way to see what is going on is to consider the 2×2 case and, in addition, to go back for a moment to the "usual" dummy indices, i and k. In this case, the right-hand side of the last equation reads

$$\sum_{i=1}^{2} \sum_{k=1}^{2} x_{i,1} x_{k,2} f(\mathbf{e}_i, \mathbf{e}_k),$$

which is a convenient stand-in for

$$\sum_{i=1}^{2} x_{i,1} \left[\sum_{k=1}^{2} x_{k,2} f(\mathbf{e}_i, \mathbf{e}_k) \right].$$

Returning now to the general case, we observe that all the terms in the multiple sum for which $p(i) = p(j)$ for some i and j, $i \neq j$, are equal to zero. For, let us take a typical term in the multiple sum:

$$x_{p(1),1} x_{p(2),2} \ldots x_{p(n),n} f(\mathbf{e}_{p(1)}, \ldots, \mathbf{e}_{p(i)}, \ldots, \mathbf{e}_{p(j)}, \ldots, \mathbf{e}_{p(n)}).$$

If $p(i) = p(j)$, then the n-tuple $\langle \mathbf{e}_{p(1)}, \ldots, \mathbf{e}_{p(i)}, \ldots, \mathbf{e}_{p(j)}, \ldots, \mathbf{e}_{p(n)} \rangle$ contains the same vector twice and, by Proposition 14.1, the value of f at this n-tuple must be zero. We see, then, that the only terms in the multiple sum that are not known *a priori* to be zero are terms of the form

$$x_{p(1),1} x_{p(2),2} \ldots x_{p(n),n} f(\mathbf{e}_{p(1)}, \ldots, \mathbf{e}_{p(n)})$$

for which $\{ p(1), p(2), \ldots, p(n) \} = \{1, 2, \ldots, n\}$. In other words, a given term in the multiple sum is not known *a priori* to be zero iff $p(1), p(2), \ldots, p(n)$ are the *values of a permutation*, call it p, of the integers $1, 2, \ldots, n$. Taking account of this fact, we may now write $f(\mathbf{X})$ as follows:

$$f(\mathbf{X}) = \sum_{p \in \mathscr{P}} x_{p(1),1} x_{p(2),2} \ldots x_{p(n),n} f(\mathbf{e}_{p(1)}, \ldots, \mathbf{e}_{p(n)}),$$

where $\mathscr{P}$ is defined to be the set of all permutations of the integers $1, 2, \ldots, n$, so that the symbol $\sum_{p \in \mathscr{P}}$ means that we are taking the sum of the terms

$$x_{p(1),1} x_{p(2),2} \ldots x_{p(n),n} f(\mathbf{e}_{p(1)}, \ldots, \mathbf{e}_{p(n)})$$

over all the permutations p of the integers $1, 2, \ldots, n$.

Now, f is an *alternating* n-linear form, and therefore

$$f(\mathbf{e}_{p(1)}, \ldots, \mathbf{e}_{p(n)}) = (sgn\ p) f(\mathbf{e}_1, \ldots, \mathbf{e}_n)$$
$$= (sgn\ p) f(\mathbf{I}),$$

where $\mathbf{I}$ is the $n \times n$ identity matrix. Thus, our final formula is

$$f(\mathbf{X}) = f(\mathbf{I}) \sum_{p \in \mathscr{P}} (sgn\ p) x_{p(1),1} x_{p(2),2} \ldots x_{p(n),n},$$

where, once again, the summation runs over all the permutations p of the integers $1, 2, \ldots, n$.

The quantity

$$\sum_{p \in \mathscr{P}} (sgn\, p) x_{p(1),1} x_{p(2),2} \cdots x_{p(n),n}$$

is called the **determinant** of the matrix $\mathbf{X}$, and it is denoted $det\, \mathbf{X}$. The function which associates every real $n \times n$ matrix with its determinant is referred to as the **determinant function**.

Let us state the result of the foregoing argument formally:

Proposition 14.2

If f is an alternating n-linear form in $\mathbf{R}^n$ and if $\mathbf{X}$ is a real $n \times n$ matrix, then

$$f(\mathbf{X}) = \lambda\, det\, \mathbf{X},$$

where λ, a real number that does not depend upon $\mathbf{X}$, is given by $\lambda = f(\mathbf{I})$.

PROOF:
See above. ∎

It is easy to verify that the determinant function is itself an n-linear form in $\mathbf{R}^n$. In fact, the determinant function is the unique alternating n-linear form in $\mathbf{R}^n$ whose value at $\mathbf{I}$ is 1. This assertion does not follow from Proposition 14.2, because it is conceivable that the only alternating n-linear form is the function that assigns the value 0 to every real $n \times n$ matrix. Thus, we must resort to direct calculation in order to exhibit the fact that the determinant function is an alternating n-linear form. This calculation runs as follows: Let $\mathbf{X}$ be a real $n \times n$ matrix with columns $\mathbf{x}_1, \ldots, \mathbf{x}_n$. Also, let $\hat{\mathbf{X}}$ be the matrix obtained when the j-th column, $\mathbf{x}_j$, of $\mathbf{X}$ is replaced by some vector $\hat{\mathbf{x}}_j$ in $\mathbf{R}^n$. Finally, let $\hat{\hat{\mathbf{X}}}$ be the matrix obtained when the j-th column of $\mathbf{X}$ is replaced by the *sum* $\mathbf{x}_j + \hat{\mathbf{x}}_j$. Then,

$$\begin{aligned}
det\, \hat{\hat{\mathbf{X}}} &= \sum_{p \in \mathscr{P}} (sgn\, p) x_{p(1),1} \cdots (x_{p(j),j} + \hat{x}_{p(j),j}) \cdots x_{p(n),n} \\
&= \sum_{p \in \mathscr{P}} (sgn\, p) x_{p(1),1} \cdots x_{p(j),j} \cdots x_{p(n),n} \\
&\quad + \sum_{p \in \mathscr{P}} (sgn\, p) x_{p(1),1} \cdots \hat{x}_{p(j),j} \cdots x_{p(n),n} \\
&= det\, \mathbf{X} + det\, \hat{\mathbf{X}}.
\end{aligned}$$

Similarly, if $\mathbf{X}^{\alpha}$ is the matrix obtained when the j-th column of $\mathbf{X}$, $\mathbf{x}_j$, is replaced by $\alpha \mathbf{x}_j$, where α is some real number, then

$$\begin{aligned}
det\, \mathbf{X}^{\alpha} &= \sum_{p \in \mathscr{P}} (sgn\, p) x_{p(1),1} \cdots \alpha x_{p(j),j} \cdots x_{p(n),n} \\
&= \alpha\, det\, \mathbf{X}.
\end{aligned}$$

So the determinant function is, indeed, an n-linear form in $\mathbf{R}^n$. To see that the determinant function is *alternating*, consider an arbitrary permutation q of the integers $1, 2, \ldots, n$. We must show that

$$det\,(\mathbf{x}_{q(1)}, x_{q(2)}, \ldots, \mathbf{x}_{q(n)}) = (sgn\,q)\,det\,\mathbf{X}.$$

Proceeding formally, we may write

$$det\,(\mathbf{x}_{q(1)}, \ldots, \mathbf{x}_{q(n)}) = \sum_{p \in \mathscr{P}} (sgn\,p)\,x_{p(1),q(1)} \cdots x_{p(n),q(n)}.$$

Note that if $q(j) = k$, then

$$x_{p(j),q(j)} = x_{p(j),k} = x_{p(q^{-1}(k)),k}.$$

Let $\hat{p}$ be the composition of q^{-1} and p:

$$\hat{p} = p \circ q^{-1},$$

and observe that, by Proposition 13.5, $sgn\,\hat{p} = (sgn\,p)/(sgn\,q)$. To make a long story short, we now have

$$det\,(\mathbf{x}_{q(1)}, \ldots, \mathbf{x}_{q(n)}) = \sum_{p \in \mathscr{P}} (sgn\,p)\,x_{\hat{p}(1),1} \cdots x_{\hat{p}(n),n}.$$

Replacing $sgn\,p$ by $(sgn\,\hat{p})(sgn\,q)$, we finally get

$$det\,(\mathbf{x}_{q(1)}, \ldots, \mathbf{x}_{q(n)}) = (sgn\,q) \sum_{\hat{p} \in \mathscr{P}} (sgn\,\hat{p})\,x_{\hat{p}(1),1} \cdots x_{\hat{p}(n),n}$$
$$= (sgn\,q)\,det\,\mathbf{X}.$$

Note that in the last step we have replaced the index of summation p by $\hat{p}$. This is permissible because the set of all permutations of the integers $1, 2, \ldots, n$ and the set of all compositions of permutations of the integers $1, 2, \ldots, n$ with the permutation q^{-1} are obviously the same set.

The fact that $det\,\mathbf{I} = 1$ may be arrived at either by direct computation or by appealing to Proposition 14.2 and using the fact that the determinant function is an alternating n-linear form.

Proposition 14.3

Let $\mathbf{D}$ be a real $n \times n$ matrix with entries d_{ij}. Assume that $d_{ij} = 0$ if $i \neq j$, i.e., that $\mathbf{D}$ is a diagonal matrix. Then,

$$det\,\mathbf{D} = d_{11}d_{22} \cdots d_{nn}.$$

PROOF:

Left to the reader. ∎

At the beginning of this chapter, we agreed, as a matter of convention, to identify an n-tuple like $\langle \mathbf{x}_1, \mathbf{x}_2, \ldots, \mathbf{x}_n \rangle$, where the $\mathbf{x}$'s are members of $\mathbf{R}^n$, with the matrix whose columns (rather than rows) are these $\mathbf{x}$'s. One of the implications of our next proposition (and its corollary) is that, in the present context, adopting the other convention would have led to precisely the same results.

Proposition 14.4

Let $\mathbf{A}$ be a real $n \times n$ matrix with entries $a_{i,j}$. Then,

$$det\ \mathbf{A}' = det\ \mathbf{A}.$$

PROOF:

By definition,

$$det\ \mathbf{A}' = \sum_{p \in \mathscr{P}} (sgn\ p) a_{1, p(1)} a_{2, p(2)} \cdots a_{n, p(n)}.$$

Consider the product

$$a_{1, p(1)} a_{2, p(2)} \cdots a_{n, p(n)}.$$

Clearly, we may change the order in which the factors $a_{j, p(j)}$ appear in this product without affecting the result. In particular, we may write

$$a_{1, p(1)} a_{2, p(2)} \cdots a_{n, p(n)} = a_{p^{-1}(1), 1} a_{p^{-1}(2), 2} \cdots a_{p^{-1}(n), n}$$

where p^{-1} is, of course, the inverse of the permutation p. Thus,

$$det\ \mathbf{A}' = \sum_{p \in \mathscr{P}} (sgn\ p) a_{p^{-1}(1), 1} \cdots a_{p^{-1}(n), n}.$$

Now, by Proposition 13.5, $sgn\ p = sgn\ p^{-1}$. Furthermore, the set of all permutations of the integers $1, 2, \ldots, n$ and the set of all inverses of permutations of the integers $1, 2, \ldots, n$ are clearly one and the same set. Hence,

$$det\ \mathbf{A}' = \sum_{q \in \mathscr{P}} (sgn\ q) a_{q(1), 1} a_{q(2), 2} \cdots a_{q(n), n},$$

where we have replaced p^{-1} by q. This last equation says that $det\ \mathbf{A}' = det\ \mathbf{A}$, as asserted. ∎

Corollary

If f is an alternating n-linear form in $\mathbf{R}^n$*, then*

$$f(\mathbf{A}') = f(\mathbf{A})$$

for every real n × n matrix $\mathbf{A}$*.*

PROOF:
Left to the reader. ∎

Let $\mathbf{A}$ and $\mathbf{B}$ be two real $n \times n$ matrices. Can anything be said about *det* $\mathbf{AB}$ in terms of *det* $\mathbf{A}$ and *det* $\mathbf{B}$? The following proposition will be instrumental in answering this question.

Proposition 14.5

Let $\mathbf{A}$ *be a fixed real n × n matrix. Then, the function f, defined by*

$$f(\mathbf{X}) = det\ \mathbf{AX}$$

for every real n × n matrix $\mathbf{X}$*, is an alternating n-linear form in* $\mathbf{R}^n$*.*

PROOF:
Let the columns of $\mathbf{X}$ be denoted $\mathbf{x}_1, \mathbf{x}_2, \ldots, \mathbf{x}_n$. Then,

$$\mathbf{AX} = \langle \mathbf{Ax}_1, \mathbf{Ax}_2, \ldots, \mathbf{Ax}_n \rangle.$$

To show that f is n-linear, we write

$$det\ (\mathbf{Ax}_1, \ldots, \mathbf{A}(\mathbf{x}_j + \hat{\mathbf{x}}_j), \ldots, \mathbf{Ax}_n)$$
$$= det\ (\mathbf{Ax}_1, \ldots, \mathbf{Ax}_j + \mathbf{A}\hat{\mathbf{x}}_j, \ldots, \mathbf{Ax}_n)$$
$$= det\ (\mathbf{Ax}_1, \ldots, \mathbf{Ax}_j, \ldots, \mathbf{Ax}_n) + det\ (\mathbf{Ax}_1, \ldots, \mathbf{A}\hat{\mathbf{x}}_j, \ldots, \mathbf{Ax}_n),$$

and also

$$det(\mathbf{Ax}_1, \ldots, \mathbf{A}(\alpha\mathbf{x}_j), \ldots, \mathbf{Ax}_n)$$
$$= det\ (\mathbf{Ax}_1, \ldots, \alpha\mathbf{Ax}_j, \ldots, \mathbf{Ax}_n)$$
$$= \alpha\ det\ (\mathbf{Ax}_1, \ldots, \mathbf{Ax}_j, \ldots, \mathbf{Ax}_n).$$

In both cases we have made use of the fact that the determinant function is an n-linear form. To see that f is alternating, consider an arbitrary permutation p of the integers $1, 2, \ldots, n$, and let $\mathbf{P}$ be its matrix. Then, $\mathbf{XP}$ is the matrix $\mathbf{X}$, with its columns rearranged according to the permutation p. We have to show that

$$det\ \mathbf{A}(\mathbf{XP}) = (sgn\ p)\ det\ \mathbf{AX}.$$

But $det\ \mathbf{A(XP)} = det\ \mathbf{(AX)P}$ and $\mathbf{(AX)P}$ is the matrix $\mathbf{AX}$ with *its* columns rearranged according to the permutation p. Using the fact that the determinant function is an alternating n-linear form, we get the desired result. ∎

Proposition 14.6

Let $\mathbf{A}$ *and* $\mathbf{B}$ *be real* $n \times n$ *matrices. Then,*

$$det\ \mathbf{AB} = (det\ \mathbf{A})(det\ \mathbf{B}).$$

PROOF:

Fix the matrix $\mathbf{A}$, and define the function f as in Proposition 14.5:

$$f(\mathbf{X}) = det\ \mathbf{AX}$$

for every real $n \times n$ matrix $\mathbf{X}$. Since f is an alternating n-linear form, we may use Proposition 14.2 and write

$$f(\mathbf{X}) = \lambda\ det\ \mathbf{X}$$

or, equivalently,

$$det\ \mathbf{AX} = \lambda\ det\ \mathbf{X}.$$

This last equation holds for every real $n \times n$ matrix $\mathbf{X}$. Furthermore, we know from Proposition 14.2 that $\lambda = f(\mathbf{I})$. Hence,

$$\lambda = det\ \mathbf{AI} = det\ \mathbf{A}.$$

Combining the last two equations, we get

$$det\ \mathbf{AX} = (det\ \mathbf{A})(det\ \mathbf{X})$$

for every real $n \times n$ matrix $\mathbf{X}$, as was to be shown. ∎

Corollary

Let $\mathbf{A}$ *and* $\mathbf{B}$ *be real* $n \times n$ *matrices. Then, the following assertions are true:*

(i) *$det\ \mathbf{AB} = det\ \mathbf{BA}$.*
(ii) *If* $\mathbf{A}$ *is nonsingular, then* $det\ \mathbf{A} \neq 0$ *and* $det\ \mathbf{A}^{-1} = (det\ \mathbf{A})^{-1}$.
(iii) *If* $\mathbf{A}$ *and* $\mathbf{B}$ *are similar, then* $det\ \mathbf{A} = det\ \mathbf{B}$.

PROOF:

Left to the reader. ∎

At the beginning of this chapter, we found (corollary to Proposition 14.1) that if A is a singular $n \times n$ matrix and f is an alternating n-linear form in $\mathbf{R}^n$, then $f(A) = 0$. Part (ii) of the foregoing corollary shows that, at least so far as the determinant function is concerned, the converse of this assertion is also true. Is this converse true for any alternating n-linear form in $\mathbf{R}^n$? Clearly not. For consider the function f defined by

$$f(\mathbf{X}) = 0$$

for every real $n \times n$ matrix $\mathbf{X}$. This function is clearly an alternating n-linear form in $\mathbf{R}^n$ and, obviously, $f(A) = 0$ does not imply that A is singular. However, if we exclude this trivial case, we get the following result:

Proposition 14.7
If f is an alternating n-linear form in $\mathbf{R}^n$ whose values are not all zero, then $f(A) = 0$ implies that A is a singular matrix.

PROOF:
Left to the reader. ∎

Our next task is to show that the determinant of a real $n \times n$ matrix can be written as a linear combination of n determinants of certain real $(n - 1) \times (n - 1)$ matrices. Let A be a real $n \times n$ matrix. We shall use the symbol A_{ij} to denote the real $(n - 1) \times (n - 1)$ matrix obtained when the i-th row and the j-th column of A are deleted. Using this notation, we now define the **cofactor of the ij-th entry** of A, or simply the **ij-th cofactor** of A, to be written $c_{ij}(A)$, as follows:

$$c_{ij}(A) = (-1)^{i+j}\, det\, A_{ij}.$$

Proposition 14.8
Let A be a real $n \times n$ matrix with entries a_{ij}. For every member i of the set $\{1, 2, \ldots, n\}$, it is true that

$$det\, A = \sum_{j=1}^{n} a_{ij} c_{ij}(A).$$

PROOF:
Let the integer i be fixed throughout the discussion, and define the real-valued function f by

$$f(\mathbf{X}) = \sum_{j=1}^{n} x_{ij} c_{ij}(\mathbf{X})$$
$$= \sum_{j=1}^{n} x_{ij}[(-1)^{i+j}]\, det\, \mathbf{X}_{ij},$$

where X is any real $n \times n$ matrix, with entries x_{ij}. We shall now show that f is an alternating n-linear form.

Let the columns of the matrix X be denoted $x_1, x_2, \ldots, x_n$, and let X^α be the matrix obtained when the k-th column of X, x_k, is replaced by αx_k, where α is some real number. We have to show that $f(X^\alpha) = \alpha f(X)$. Writing out $f(X^\alpha)$, we get

$$f(X^\alpha) = \sum_{j \neq k} x_{ij}[(-1)^{i+j}] \, det \, X_{ij}^\alpha + \alpha x_{ik}[(-1)^{i+k}] \, det \, X_{ik}.$$

The fact that the determinant function is an n-linear form implies, for $j \neq k$, that $det \, X_{ij}^\alpha = \alpha \, det \, X_{ij}$. Hence,

$$f(X^\alpha) = \alpha \sum_{j=1}^{n} x_{ij}[(-1)^{i+j}] \, det \, X_{ij}$$
$$= \alpha f(X),$$

as was to be shown.

Next, let $\hat{X}$ be the matrix obtained when the k-th column of X, x_k, is replaced by some vector $\hat{x}_k$. Furthermore, let $\hat{\hat{X}}$ be the matrix obtained when the k-th column of X is replaced by the *sum* $x_k + \hat{x}_k$. We now have to show that $f(\hat{\hat{X}}) = f(X) + f(\hat{X})$. Writing out $f(\hat{\hat{X}})$, we get

$$f(\hat{\hat{X}}) = \sum_{j \neq k} x_{ij}[(-1)^{i+j}] \, det \, \hat{\hat{X}}_{ij} + (x_{ik} + \hat{x}_{ik})[(-1)^{i+k}] \, det \, X_{ik}.$$

Once again, using the fact that the determinant function is an n-linear form, we may write, for $j \neq k$,

$$det \, \hat{\hat{X}}_{ij} = det \, X_{ij} + det \, \hat{X}_{ij}.$$

Also, note that (for the case $j = k$) $det \, X_{ik}$ and $det \, \hat{X}_{ik}$ are one and the same thing, because the k-th column is being deleted anyway. Using these facts, we arrive immediately at the desired equation:

$$f(\hat{\hat{X}}) = f(X) + f(\hat{X}).$$

Thus, f is an n-linear form in $\mathbf{R}^n$. We must now show that f is alternating.

Let p be an arbitrary permutation of the integers $1, 2, \ldots, n$, and let P be its matrix. The assertion that f is alternating may be stated as follows:

$$f(XP) = (sgn \, p) f(X).$$

Writing out $f(\mathbf{XP})$, we get†

$$f(\mathbf{XP}) = \sum_{j=1}^{n} x_{i,\,p(j)}[(-1)^{i+j}]\,det\,(\mathbf{XP})_{ij},$$

where $(\mathbf{XP})_{ij}$ is the matrix obtained when the i-th row and the j-th column of the matrix $\mathbf{XP}$ are deleted. In terms of what is happening to the matrix $\mathbf{X}$, the notation $(\mathbf{XP})_{ij}$ indicates "first permute, then delete." Obviously, the same result may also be obtained with a "first delete, then permute" procedure. Indeed, it is easy to verify that

$$(\mathbf{XP})_{ij} = \mathbf{X}_{i,\,p(j)}\,\mathbf{P}_{p(j),\,j}.$$

Thus,

$$f(\mathbf{XP}) = \sum_{j=1}^{n} x_{i,\,p(j)}[(-1)^{i+j}]\,det\,(\mathbf{X}_{i,\,p(j)}\mathbf{P}_{p(j),\,j}).$$

Now, the determinant function is an alternating n-linear form, so that $det\,(\mathbf{X}_{i,\,p(j)}\mathbf{P}_{p(j),\,j})$ is equal to $det\,\mathbf{X}_{i,\,p(j)}$ multiplied by the sign of the permutation whose matrix is $\mathbf{P}_{p(j),\,j}$. By Proposition 13.7, this sign is given by

$$[(-1)^{p(j)+j}]\,sgn\,p.$$

Hence,

$$f(\mathbf{XP}) = (sgn\,p) \sum_{j=1}^{n} x_{i,\,p(j)}[(-1)^{i+j}][(-1)^{p(j)+j}]\,det\,\mathbf{X}_{i,\,p(j)}$$

$$= (sgn\,p) \sum_{j=1}^{n} x_{i,\,p(j)}[(-1)^{i+p(j)}]\,det\,\mathbf{X}_{i,\,p(j)}.$$

We may now take the sum of the terms in this last expression in a different order, writing the term for which $p(j) = 1$ first, the term for which $p(j) = 2$ second, and so on. The result will be

$$f(\mathbf{XP}) = (sgn\,p) \sum_{k=1}^{n} x_{ik}[(-1)^{i+k}]\,det\,\mathbf{X}_{ik}$$

$$= (sgn\,p)f(\mathbf{X}),$$

as asserted.

Armed with the fact that f is an alternating n-linear form, we now

---

† Where necessary, we shall insert a comma between the two subscripts of a double-subscripted symbol.

resort to Proposition 14.2, which tells us that for every real $n \times n$ matrix **X**,

$$f(\mathbf{X}) = f(\mathbf{I}) \, det \, \mathbf{X}.$$

To complete the proof, all that has to be shown is $f(\mathbf{I}) = 1$. But this is immediate. ∎

Corollary
*Let **A** be a real $n \times n$ matrix with entries a_{ij}. For every member j of the set $\{1, 2, \ldots, n\}$ it is true that*

$$det \, \mathbf{A} = \sum_{i=1}^{n} a_{ij} c_{ij}(\mathbf{A}).$$

PROOF:
Use the foregoing proposition on $det \, \mathbf{A}'$, and then appeal to Proposition 14.4, which asserts that $det \, \mathbf{A}' = det \, \mathbf{A}$. ∎

The formulas

$$det \, \mathbf{A} = \sum_{j=1}^{n} a_{ij} c_{ij}(\mathbf{A})$$

and

$$det \, \mathbf{A} = \sum_{i=1}^{n} a_{ij} c_{ij}(\mathbf{A})$$

are called **cofactor expansions** of $det \, \mathbf{A}$. The former is referred to as the expansion of $det \, \mathbf{A}$ by the cofactors of the i-th row of **A**, and the latter is referred to as the expansion of $det \, \mathbf{A}$ by the cofactors of the j-th column of **A**.

It is interesting to note that if we write the expansion of $det \, \mathbf{A}$ by the cofactors of, say, the i-th row but use in it the cofactors of another row (these are called **alien** cofactors), the result is always zero.

Proposition 14.9
*Let **A** be a real $n \times n$ matrix with entries a_{ij}. If i and k are different members of the set $\{1, 2, \ldots, n\}$, then*

$$\sum_{j=1}^{n} a_{ij} c_{kj}(\mathbf{A}) = 0.$$

PROOF:

By Proposition 14.8, the sum

$$\sum_{j=1}^{n} a_{ij} c_{kj}(\mathbf{A})$$

is equal to the determinant of the matrix obtained from $\mathbf{A}$ by deleting the k-th row from it and inserting the i-th row in its place. In other words, this sum is equal to the determinant of a matrix that has two identical rows, and it must therefore be equal to zero. ∎

Needless to say, it is also true that if j and k are two different members of the set $\{1, 2, \ldots, n\}$, then

$$\sum_{i=1}^{n} a_{ij} c_{ik}(\mathbf{A}) = 0.$$

Consider a real $n \times n$ matrix $\mathbf{A}$, and let $\mathbf{C}(\mathbf{A})$ be the real $n \times n$ matrix whose ij-th entry is the ij-th cofactor of $\mathbf{A}$, $c_{ij}(\mathbf{A})$. The transpose of $\mathbf{C}(\mathbf{A})$ is referred to as the **adjoint** of $\mathbf{A}$, and it is denoted $adj\ \mathbf{A}$:

$$adj\ \mathbf{A} = \mathbf{C}(\mathbf{A})'.$$

If $\mathbf{A}$ happens to be nonsingular, then $adj\ \mathbf{A}$ and $\mathbf{A}^{-1}$ are simply multiples of each other.

Proposition 14.10

Let $\mathbf{A}$ be a real $n \times n$ matrix. If $\mathbf{A}$ is nonsingular, then

$$\mathbf{A}^{-1} = \left(\frac{1}{det\ \mathbf{A}}\right) adj\ \mathbf{A}.$$

PROOF:

The findings of Proposition 14.8 and Proposition 14.9 may be combined to yield the following equation:

$$\mathbf{A}(adj\ \mathbf{A}) = (det\ \mathbf{A})\mathbf{I}.$$

This equation is true whether $\mathbf{A}$ is singular or not. However, if $\mathbf{A}$ is nonsingular, then we can premultiply both sides by $\mathbf{A}^{-1}$ and multiply both sides by the real number $1/det\ \mathbf{A}$, to get the desired result. ∎

Corollary (Cramer's Rule)

Consider a real $n \times n$ matrix $\mathbf{A}$ and a vector $\mathbf{b}$ in $\mathbf{R}^n$. Let the symbol $\mathbf{A}(\mathbf{b}, j)$ stand for the matrix obtained when we delete the j-th column of $\mathbf{A}$ and insert the vector $\mathbf{b}$ in its place. Assume that the matrix $\mathbf{A}$ is nonsingular, and define x_j^* as follows:

$$x_j^* = \frac{det\ \mathbf{A}(\mathbf{b}, j)}{det\ \mathbf{A}}$$

for $j = 1, 2, \ldots, n$. Let $\mathbf{x}^*$ be the element of $\mathbf{R}^n$ whose components are these x_j^*'s.

$$\mathbf{x}^* = \begin{bmatrix} x_1^* \\ x_2^* \\ \cdot \\ \cdot \\ \cdot \\ x_n^* \end{bmatrix}.$$

Then, $\mathbf{x}^*$ is the (unique) solution of the equation $\mathbf{Ax} = \mathbf{b}$.

PROOF:

The solution of the equation $\mathbf{Ax} = \mathbf{b}$ is given by $\mathbf{A}^{-1}\mathbf{b}$. In view of the foregoing proposition, this solution may also be written

$$\left(\frac{1}{det\ \mathbf{A}}\right)(adj\ \mathbf{A})\mathbf{b}.$$

Writing out the j-th component of the vector $(adj\ \mathbf{A})\mathbf{b}$ and the expansion of $det\ \mathbf{A}(\mathbf{b}, j)$ by the cofactors of the j-th column, we find that the two formulas coincide. ∎

PROBLEMS

14.1. Consider the set of all n-linear forms in $\mathbf{R}^n$. Define addition and multiplication by a real number on this set in the natural way:

$$(f + g)(\mathbf{X}) = f(\mathbf{X}) + g(\mathbf{X})$$
$$(\lambda f)(\mathbf{X}) = \lambda f(\mathbf{X})$$

for every real $n \times n$ matrix $\mathbf{X}$. With these definitions, the set of all n-linear forms in $\mathbf{R}^n$ becomes a real linear space. Find a basis for this space, and prove that what you have found is indeed a basis by exhibiting linear independence and spanning. (Hint: Consider functions that carry a real $n \times n'$ matrix $\mathbf{X}$ into the product of n entries from $\mathbf{X}$, one entry from each column.) What is the dimension of the space of all n-linear forms in $\mathbf{R}^n$? Is the set of all *alternating* n-linear forms in $\mathbf{R}^n$ a subspace of this real linear space? If so, what is the dimension of this subspace?

14.2. In the text, the function f defined by

$$f(\mathbf{X}) = \xi_1 \xi_2 \ldots \xi_n,$$

where ξ_j is the sum of the entries in the j-th column of $\mathbf{X}$, was mentioned as an example of an n-linear form in $\mathbf{R}^n$. Express this function in terms of the basis you have constructed in Problem 14.1.

14.3. Consider two distinct members, i and j, of the set $\{1, 2, \ldots, n\}$ and let $\mathbf{E}$ be the elementary matrix obtained when the ij-th entry of the $n \times n$ identity matrix $\mathbf{I}$ is changed from 0 to α. Show that if f is an alternating n-linear form in $\mathbf{R}^n$, then

$$f(\mathbf{EX}) = f(\mathbf{XE}) = f(\mathbf{X})$$

for every real $n \times n$ matrix $\mathbf{X}$.

14.4. Prove Proposition 14.7.

14.5. Show that the determinant of an orthogonal matrix is either -1 or 1.

14.6. Let the following $n \times n$ matrix

$$\begin{bmatrix} 1 & \alpha & 0 & 0 & \ldots & 0 & 0 \\ \alpha & 1 & \alpha & 0 & \ldots & 0 & 0 \\ 0 & \alpha & 1 & \alpha & \ldots & 0 & 0 \\ 0 & 0 & \alpha & 1 & \ldots & 0 & 0 \\ \cdot & \cdot & \cdot & \cdot & & \cdot & \cdot \\ \cdot & \cdot & \cdot & \cdot & & \cdot & \cdot \\ \cdot & \cdot & \cdot & \cdot & & \cdot & \cdot \\ 0 & 0 & 0 & 0 & \ldots & \alpha & 1 \end{bmatrix}$$

be denoted $\mathbf{A}_n$. Thus, for example, $\mathbf{A}_{n-1}$ is the matrix that has the same form (with the same value of the real number α) with only $n-1$ rows and columns. Let $det\, \mathbf{A}_n$ be denoted D_n. Prove that D_n satisfies the following recursive relationship:

$$D_n = D_{n-1} - \alpha^2 D_{n-2} \quad \text{for} \quad n = 2, 3, \ldots$$
$$D_0 = D_1 = 1.$$

Readers familiar with the binomial coefficients should also try to prove that D_n is given explicitly by

$$D_{2m} = \sum_{k=0}^{m} \binom{2m-k}{k}(-1)^k \alpha^{2k}$$

$$D_{2m+1} = \sum_{k=0}^{m} \binom{2m-k+1}{k}(-1)^k \alpha^{2k}.$$

(7) Solve the following system of equations, using Cramer's Rule and cofactor expansions:

$$x_1 + x_5 = 1$$
$$2x_1 + 3x_2 + x_3 + x_4 = 0$$
$$x_2 + x_3 = 0$$
$$x_1 - x_3 + 2x_4 = 0.$$

15 Characteristic Vectors

and Characteristic Roots

In this chapter, we shall be concerned with the following question: If T is a linear transformation on $\mathbf{R}^n$ to $\mathbf{R}^n$, is there a basis $\mathbf{B}$ for $\mathbf{R}^n$ such that the matrix of T relative to $\mathbf{B}$ is a diagonal matrix? If the answer to this question is in the affirmative, then the linear transformation T is said to be **diagonalizable**. It is, of course, possible to ask the foregoing question in terms of the natural matrix of the linear transformation T, rather than in terms of T itself: If $\mathbf{A}$ is a real $n \times n$ matrix, is there a real diagonal matrix $\mathbf{D}$ such that $\mathbf{A}$ and $\mathbf{D}$ are similar? If the answer to this question is in the affirmative, then we say, in analogy with the previous definition, that the matrix $\mathbf{A}$ is diagonalizable.

Let $\mathbf{A}$ be a real $n \times n$ matrix and assume that $\mathbf{A}$ is diagonalizable. By definition, this means that there exists a nonsingular matrix $\mathbf{C}$ such that

$$\mathbf{C}^{-1}\mathbf{A}\mathbf{C} = \mathbf{D},$$

or, equivalently, such that

$$\mathbf{A}\mathbf{C} = \mathbf{C}\mathbf{D},$$

where $\mathbf{D}$ is a diagonal $n \times n$ matrix. This last equation can be broken down into n separate equations in the following manner: Let $\lambda_1, \lambda_2, \ldots, \lambda_n$ be the entries on the main diagonal of $\mathbf{D}$, and let $\mathbf{c}_1, \mathbf{c}_2, \ldots, \mathbf{c}_n$ be the columns of $\mathbf{C}$. Then, the statement $\mathbf{AC} = \mathbf{CD}$ is equivalent to

$$\mathbf{Ac}_j = \lambda_j \mathbf{c}_j \quad \text{for} \quad j = 1, 2, \ldots, n.$$

In other words, $\mathbf{A}$ is diagonalizable iff there exist n linearly independent elements, $\mathbf{c}_1, \mathbf{c}_2, \ldots, \mathbf{c}_n$, of $\mathbf{R}^n$, and n real numbers, $\lambda_1, \lambda_2, \ldots, \lambda_n$, such that the foregoing equation holds. This brings us to the notions of a characteristic vector and a characteristic root.

Once again, let $\mathbf{A}$ be an arbitrary real $n \times n$ matrix. A member $\mathbf{x}$ of $\mathbf{R}^n$ is said to be a **characteristic vector** (or an **eigenvector**) of $\mathbf{A}$ iff $\mathbf{x} \neq \mathbf{0}$ and there exists a real number λ such that

$$\mathbf{Ax} = \lambda \mathbf{x}.$$

If such a vector $\mathbf{x}$ exists, then the real number λ appearing in the definition is referred to as a **characteristic root** (or as an **eigenvalue**) of the matrix $\mathbf{A}$. We say, furthermore, that $\mathbf{x}$ is a characteristic vector **associated with** the characteristic root λ. The question of the diagonalizability of a matrix $\mathbf{A}$ is equivalent to the question of the existence of n linearly independent characteristic vectors of $\mathbf{A}$.

The notions we have just introduced have applications in a great many areas. The following is a rather simpleminded example of the use of characteristic roots in the study of a model based on a difference equation. Let $\mathscr{S}$ be a system of some sort, say an economic system, and assume that the state of the system $\mathscr{S}$ at time t is described by an element $\mathbf{x}_t$ of $\mathbf{R}^n$. Suppose further that the evolution of the system through time is given by the following law:

$$\mathbf{x}_t = \mathbf{A}\mathbf{x}_{t-1}, \quad t = 1, 2, \ldots$$

where $\mathbf{A}$ is a real $n \times n$ matrix. One defines an *equilibrium state* of the system $\mathscr{S}$ as an element $\mathbf{x}^*$ of $\mathbf{R}^n$ with the property that $\mathbf{A}\mathbf{x}^* = \mathbf{x}^*$. Clearly, the null element $\mathbf{0}$ is always an equilibrium state. Now suppose that $\mathbf{A}$ is diagonalizable. Then, it is possible to derive statements about the long-run behavior of the system $\mathscr{S}$ by studying the characteristic roots, $\lambda_1, \lambda_2, \ldots, \lambda_n$, of $\mathbf{A}$. For example, if these roots satisfy the condition $-1 < \lambda_j < 1$ for $j = 1, 2, \ldots,$ n, then the system $\mathscr{S}$ will, in the long run, approach the equilibrium state $\mathbf{0}$, regardless of the initial state $\mathbf{x}_0$. This last sentence is deficient in that the word "approach" has not been defined. We shall leave this matter to the reader's intuition.

Let $\mathbf{A}$ be a real $n \times n$ matrix, and suppose that we are told that $\mathbf{A}$ has a characteristic root. How shall we go about finding this root? The answer lies in the following proposition.

Proposition 15.1
A real number λ is a characteristic root of a real $n \times n$ matrix $\mathbf{A}$ iff det $(\mathbf{A} - \lambda\mathbf{I})$ $= 0$, that is, iff the matrix $\mathbf{A} - \lambda\mathbf{I}$ is singular. (The symbol $\mathbf{I}$ stands, as usual, for the $n \times n$ identity matrix.)

PROOF:
By definition, λ is a characteristic root of $\mathbf{A}$ iff

$$\mathbf{A}\mathbf{x} = \lambda\mathbf{x}$$

for some nonnull member $\mathbf{x}$ of $\mathbf{R}^n$. This last equation may be written $\mathbf{A}\mathbf{x} = \lambda\mathbf{I}\mathbf{x}$, which in turn is equivalent to

$$(\mathbf{A} - \lambda\mathbf{I})\mathbf{x} = \mathbf{0}.$$

We know that the vector $\mathbf{x}$ is a nontrivial solution of this homogeneous linear equation, and such a nontrivial solution exists iff the matrix $\mathbf{A} - \lambda\mathbf{I}$ has a nontrivial null space, i.e., iff $\mathbf{A} - \lambda\mathbf{I}$ is singular. ∎

Thus, if $\mathbf{A}$ is a real $n \times n$ matrix and if $\mathbf{A}$ has a characteristic root, then this characteristic root must be a solution of the following equation in λ:

$$det\,(\mathbf{A} - \lambda\mathbf{I}) = 0.$$

In fact, Proposition 15.1 tells us that converse of this assertion is also true. Every solution of the equation $det\,(\mathbf{A} - \lambda\mathbf{I}) = 0$ is a characteristic root of $\mathbf{A}$.

Let f be a function on $\mathbf{R}$ into $\mathbf{R}$. We say that f is a (real) **polynomial of degree n** iff for every $x \in \mathbf{R}$,

$$f(x) = a_0 + a_1x + a_2x^2 + \cdots + a_nx^n$$

for some real numbers $a_0, a_1, \ldots, a_n$, with $a_n \neq 0$. If f is a polynomial degree n and x_0 is a real number satisfying

$$f(x_0) = 0,$$

then x_0 is referred to as a **root** of the polynomial f.

Let us return now to the expression $det\,(\mathbf{A} - \lambda\mathbf{I})$. The function that associates each real number λ with the real number $det\,(\mathbf{A} - \lambda\mathbf{I})$ is clearly a polynomial of degree n. [To see this, use the definition of the determinant to write out $det\,(\mathbf{A} - \lambda\mathbf{I})$.] This polynomial is referred to as the **characteristic polynomial** of $\mathbf{A}$. The characteristic roots of $\mathbf{A}$ are, of course, the roots of the characteristic polynomial of $\mathbf{A}$. One further definition: The equation $det\,(\mathbf{A} - \lambda\mathbf{I}) = 0$ is often referred to as the **characteristic equation** of $\mathbf{A}$.

We come now to the question of the existence of characteristic roots and characteristic vectors. Given a real $n \times n$ matrix $\mathbf{A}$, is there a nonnull vector $\mathbf{x}$ in $\mathbf{R}^n$ such that $\mathbf{Ax} = \lambda\mathbf{x}$ for some real number λ? The answer is not in general affirmative. Consider, for example, the matrix

$$\mathbf{A} = \begin{bmatrix} 0 & 1 \\ -1 & 0 \end{bmatrix}.$$

The characteristic polynomial, $det\,(\mathbf{A} - \lambda\mathbf{I})$, is in this case simply $\lambda^2 + 1$, and there exists no real number λ for which $\lambda^2 + 1 = 0$. Hence, $\mathbf{A}$ has no characteristic roots and, *a fortiori*, it has no characteristic vectors. The reader may be aware of the fact that the polynomial $\lambda^2 + 1$, while it has no real roots, does have complex roots. Indeed, there exists a celebrated theorem, called the Fundamental Theorem of Algebra, which states that every polynomial of degree n has a complex root. In other words, if $\mathbf{A}$ is a real $n \times n$ matrix, then there exists a complex number μ such that $det\,(\mathbf{A} - \mu\mathbf{I}) = 0$. However, this fact does not permit us to conclude that there exists an element $\mathbf{x}$ of $\mathbf{R}^n$ such that $\mathbf{Ax} = \mu\mathbf{x}$. In a real linear space, we are allowed to multiply elements only by real numbers, and μ is not in general a real number. The quantity $\mu\mathbf{x}$ is therefore not well defined. Indeed, it is clear that for the equation $\mathbf{A}\mathbf{x} = \mu\mathbf{x}$ to hold, it is in general necessary that $\mathbf{x}$ be a vector in a linear space whose members are n-tuples of *complex* numbers, and in which scalar multiplication is defined as multiplication by a complex number. In short, the notion of a real linear space is not broad enough for taking advantage of the Fundamental Theorem of Algebra. It turns out, however, that *symmetric* matrices always have real characteristic roots, so that, by restricting our attention to symmetric matrices, we shall be able to proceed with the analysis without leaving the framework of real linear spaces. Of course, this restriction of the analysis to symmetric matrices entails a considerable loss of generality, but it should be noted that the arguments used in analyzing this special case are similar in spirit to those used in the general case. Unfortunately, the proof of the assertion that a real symmetric matrix has real characteristic roots, even though it is quite simple, requires notions that lie beyond the scope of this volume. The following proposition is therefore stated without proof.

Proposition 15.2

Let $\mathbf{A}$ *be a real* $n \times n$ *matrix. If* $\mathbf{A}$ *is symmetric, then there exist* n *real numbers,* $\lambda_1, \lambda_2, \ldots, \lambda_n$, *such that*

$$det\,(\mathbf{A} - \lambda\mathbf{I}) = (\lambda_1 - \lambda)(\lambda_2 - \lambda) \cdots (\lambda_n - \lambda).$$

PROOF:

Omitted. ∎

From the existence of real characteristic roots follows the existence of characteristic vectors in $\mathbf{R}^n$. If λ_j is a characteristic root of the symmetric matrix $\mathbf{A}$, then all we have to do to find a characteristic vector associated with λ_j is to locate a nontrivial solution of the homogeneous equation $(\mathbf{A} - \lambda_j\mathbf{I})\mathbf{x} = \mathbf{0}$. Such a nontrivial solution must exist, by Proposition 15.1.

In order to show that a real symmetric $n \times n$ matrix $\mathbf{A}$ is diagonalizable, we have to find n *linearly independent* characteristic vectors of $\mathbf{A}$. As a first step in this direction, we have the following proposition, which asserts that if two characteristic vectors are associated with two different characteristic roots, then these characteristic vectors are not only linearly independent, but actually orthogonal (to each other).

Proposition 15.3

Let $\mathbf{A}$ *be a real symmetric* $n \times n$ *matrix, with characteristic roots* $\lambda_1, \lambda_2, \ldots,$ λ_n. *If* $\lambda_i \neq \lambda_j$ *for some* i *and* j, *and if* $\mathbf{x}_i$ *and* $\mathbf{x}_j$ *are, respectively, a characteristic vector associated with* λ_i *and a characteristic vector associated with* λ_j, *then* $\mathbf{x}_i$ *and* $\mathbf{x}_j$ *are orthogonal.*

PROOF:

By definition, we have

$$\mathbf{A}\mathbf{x}_i = \lambda_i \mathbf{x}_i$$
$$\mathbf{A}\mathbf{x}_j = \lambda_j \mathbf{x}_j.$$

In the first of these equations, let us take the inner product of both sides with $\mathbf{x}_j$, and in the second let us take the inner product of both sides with $\mathbf{x}_i$. The result will be

$$(\mathbf{x}_j, \mathbf{A}\mathbf{x}_i) = \lambda_i(\mathbf{x}_j, \mathbf{x}_i)$$
$$(\mathbf{x}_i, \mathbf{A}\mathbf{x}_j) = \lambda_j(\mathbf{x}_i, \mathbf{x}_j).$$

The fact that $\mathbf{A}$ is symmetric implies that

$$(\mathbf{x}_j, \mathbf{A}\mathbf{x}_i) = (\mathbf{x}_i, \mathbf{A}\mathbf{x}_j).$$

To see this, write the inner product $(\mathbf{x}_j, \mathbf{A}\mathbf{x}_i)$ in the alternative notation, $\mathbf{x}_j'\mathbf{A}\mathbf{x}_i$. Since this inner product is a real number, its transpose (when it is looked upon as a real 1×1 matrix) is surely equal to itself. In other words,

$$\mathbf{x}_j'\mathbf{A}\mathbf{x}_i = [\mathbf{x}_j'\mathbf{A}\mathbf{x}_i]' = \mathbf{x}_i'\mathbf{A}'\mathbf{x}_j = \mathbf{x}_i'\mathbf{A}\mathbf{x}_j,$$

as asserted. Therefore,

$$\lambda_i(\mathbf{x}_j, \mathbf{x}_i) = \lambda_j(\mathbf{x}_i, \mathbf{x}_j),$$

and since $(\mathbf{x}_j, \mathbf{x}_i) = (\mathbf{x}_i, \mathbf{x}_j)$ and $\lambda_i \neq \lambda_j$, it must be true that $(\mathbf{x}_i, \mathbf{x}_j) = 0$. ∎

Corollary

Let $\mathbf{A}$ *be a real symmetric* $n \times n$ *matrix, with characteristic roots* $\lambda_1, \lambda_2, \ldots,$ λ_n. *If these characteristic roots are all different from one another, then* $\mathbf{A}$ *is diagonalizable.*

PROOF:

Let $\mathbf{x}_1, \mathbf{x}_2, \ldots, \mathbf{x}_n$ be characteristic vectors of $\mathbf{A}$ that are associated, respectively, with $\lambda_1, \lambda_2, \ldots, \lambda_n$. By the foregoing proposition, the vectors $\mathbf{x}_1, \mathbf{x}_2, \ldots, \mathbf{x}_n$ are pairwise orthogonal, and therefore they must be linearly independent. If $\mathbf{X}$ is the matrix whose columns are the vectors $\mathbf{x}_1, \mathbf{x}_2, \ldots, \mathbf{x}_n$, then $\mathbf{X}^{-1}\mathbf{AX}$ is a diagonal matrix. ∎

It should be noted that if $\mathbf{x}$ is a characteristic vector of a real $n \times n$ matrix $\mathbf{A}$, and if α is any real number other than zero, then $\alpha\mathbf{x}$ is also a characteristic vector of $\mathbf{A}$. Moreover, $\mathbf{x}$ and $\alpha\mathbf{x}$ are clearly associated with the same characteristic root. (This fact is sometimes referred to by saying that characteristic vectors are determined "only up to a constant of proportionality.") In particular, if $\mathbf{x}$ is a characteristic vector of $\mathbf{A}$, then there always exists a real number α (namely, $\alpha = 1/\sqrt{(\mathbf{x}, \mathbf{x})}$) such that $\alpha\mathbf{x}$ has the property that $(\alpha\mathbf{x}, \alpha\mathbf{x}) = 1$. A characteristic vector $\mathbf{x}$ satisfying the property $(\mathbf{x}, \mathbf{x}) = 1$ is said to be **normalized**. In the foregoing corollary to Proposition 15.3, we can clearly pick the characteristic vectors $\mathbf{x}_1, \mathbf{x}_2, \ldots, \mathbf{x}_n$ to be normalized, with the result that the matrix $\mathbf{X}$, whose columns are the vectors $\mathbf{x}_1, \mathbf{x}_2, \ldots, \mathbf{x}_n$, becomes an orthogonal matrix. This brings us to the following definition. Two real $n \times n$ matrices, A and B, are said to be **orthogonally similar** iff there exists an orthogonal matrix $\mathbf{U}$ such that $\mathbf{U'AU} = \mathbf{B}$. Using this definition, we may state the corollary to Proposition 15.3 as follows: *A real symmetric matrix with distinct characteristic roots is orthogonally similar to a diagonal matrix.* The remainder of this chapter will be devoted to showing that any real symmetric matrix, even if its characteristic roots are not all distinct, is orthogonally similar to a diagonal matrix. In the course of proving this assertion, we shall have use for the following proposition.

Proposition 15.4

Let A and B be two real $n \times n$ matrices. If A and B are similar, then the characteristic polynomials of A and B coincide and, therefore, A and B have the same characteristic roots.

PROOF:

By hypothesis, there exists a nonsingular matrix $\mathbf{C}$ such that $\mathbf{C}^{-1}\mathbf{AC} = \mathbf{B}$. Thus,

$$\begin{aligned} det\,(\mathbf{B} - \lambda\mathbf{I}) &= det\,(\mathbf{C}^{-1}\mathbf{AC} - \lambda\mathbf{I}) \\ &= det\,(\mathbf{C}^{-1}\mathbf{AC} - \lambda\mathbf{C}^{-1}\mathbf{IC}) \\ &= det\,(\mathbf{C}^{-1}(\mathbf{A} - \lambda\mathbf{I})\mathbf{C}) \\ &= det\,(\mathbf{A} - \lambda\mathbf{I}). \blacksquare \end{aligned}$$

Consider a real symmetric $n \times n$ matrix $\mathbf{A}$, and suppose that $\lambda_1, \lambda_2, \ldots, \lambda_m$ are the *distinct* characteristic roots of $\mathbf{A}$. By Proposition 15.2, there exist m positive integers $k_1, k_2, \ldots, k_m$ satisfying

$$\sum_{i=1}^{m} k_i = n,$$

such that

$$det\ (\mathbf{A} - \lambda\mathbf{I}) = (\lambda_1 - \lambda)^{k_1}(\lambda_2 - \lambda)^{k_2} \cdots (\lambda_m - \lambda)^{k_m}.$$

The integers $k_1, k_2, \ldots, k_m$ are referred to as the respective **multiplicities** of the characteristic roots $\lambda_1, \lambda_2, \ldots, \lambda_m$.

Proposition 15.5

Let $\mathbf{A}$ *be a real symmetric* $n \times n$ *matrix, and let* $\lambda_1, \lambda_2, \ldots, \lambda_m$ *be the distinct characteristic roots of* $\mathbf{A}$, *with respective multiplicities* $k_1, k_2, \ldots, k_m$. *For each member* j *of the set* $\{1, 2, \ldots, m\}$ *let a set* S_j *be defined as follows:*

$$S_j = \{\mathbf{x} \,|\, \mathbf{x} \in \mathbf{R}^n \quad and \quad \mathbf{Ax} = \lambda_j\mathbf{x}\}.$$

(In other words, S_j *is the set of all characteristic vectors that are associated with the characteristic root* λ_j.) *Then, for* $j = 1, 2, \ldots, m$, S_j *is a linear subspace of* $\mathbf{R}^n$ *and, furthermore,*

$$dim\ S_j = k_j.$$

PROOF:

That S_j is a subspace is clear. In fact, S_j is nothing other than the null space of the matrix $\mathbf{A} - \lambda_j\mathbf{I}$. It is somewhat harder to show that the dimension of S_j is equal to the multiplicity of λ_j. We proceed as follows: For each of the subspaces $S_1, S_2, \ldots, S_m$, we select an orthonormal basis. This is possible, by Proposition 7.4. Let the orthonormal basis for S_j be denoted $\mathbf{B}_j$, for $j = 1, 2, \ldots, m$. The dimension of S_j is given by the number of components of $\mathbf{B}_j$. Denote this number k'_j. What we have to show is that $k'_j = k_j$ for $j = 1, 2, \ldots, m$. For convenience, let r be the sum of the integers k'_j:

$$\sum_{j=1}^{m} k'_j = r.$$

Now let us take the bases $\mathbf{B}_1, \mathbf{B}_2, \ldots, \mathbf{B}_m$ and join them together to form one r-tuple of vectors in $\mathbf{R}^n$. Let $\mathbf{B}$ be this r-tuple, $\mathbf{B} = \langle \mathbf{B}_1, \mathbf{B}_2, \ldots, \mathbf{B}_m \rangle$, and let S be the set spanned by $\mathbf{B}$. S is clearly a subspace of $\mathbf{R}^n$ and $\mathbf{B}$ is clearly an orthonormal basis for S. This latter fact is true because the components of each $\mathbf{B}_j$ are orthonormal to begin with, and if we take a component of $\mathbf{B}_j$ and a component of $\mathbf{B}_i$, for $i \neq j$, then these two components are orthogonal, by Proposition 15.3. Now let $S^\perp$ be the orthogonal complement of S. Recall that no nonnull vector can be simultaneously in S and in $S^\perp$. Select an orthonormal basis for $S^\perp$ and call it $\mathbf{B}_0$. By Proposition 7.6, $\mathbf{B}_0$ has $n - r$ components. If $S^\perp$

is trivial, i.e., if $S^{\perp} = \{0\}$, then $n - r = 0$ and $\mathbf{B}_0$ is simply the empty set. Our aim is to show that this is, indeed, the case. By the definition of $S^{\perp}$, every component of $\mathbf{B}_0$ is orthogonal to every component of $\mathbf{B}$. Let us adjoin the bases $\mathbf{B}$ and $\mathbf{B}_0$ together, and let us denote the result $\hat{\mathbf{B}}$:

$$\hat{\mathbf{B}} = \langle \mathbf{B}, \mathbf{B}_0 \rangle.$$

Clearly, we may look upon $\hat{\mathbf{B}}$ as a real $n \times n$ matrix, namely, as the matrix whose columns are the components of $\mathbf{B}_1, \mathbf{B}_2, \ldots, \mathbf{B}_m$ and $\mathbf{B}_0$. $\hat{\mathbf{B}}$ is, in fact, an orthogonal matrix.

Consider the matrix $\mathbf{C}$ defined by $\mathbf{C} = \hat{\mathbf{B}}'\mathbf{A}\hat{\mathbf{B}}$. A moment's reflection will reveal that $\mathbf{C}$ has the following form:

$$\mathbf{C} = \begin{bmatrix} \mathbf{D} & \mathbf{O} \\ \mathbf{O}' & \mathbf{C}_0 \end{bmatrix},$$

where $\mathbf{C}_0$ is some real $(n - r) \times (n - r)$ matrix, $\mathbf{O}$ is the $r \times (n - r)$ null matrix, and $\mathbf{D}$ is an $r \times r$ diagonal matrix whose diagonal entries are as follows: λ_1 repeated k_1' times, λ_2 repeated k_2' times, and so on, down to λ_m which is repeated k_m' times. (In order to see more clearly why $\mathbf{C}$ is of this form, it is perhaps worthwhile to look first at the matrix $\mathbf{A}\hat{\mathbf{B}}$, and consider it column by column.) Given that $\mathbf{C}$ is of this form, it follows that the characteristic polynomial of $\mathbf{C}$ is given by

$$det\,(\mathbf{C} - \lambda\mathbf{I}) = (\lambda_1 - \lambda)^{k_1'}(\lambda_2 - \lambda)^{k_2'} \cdots (\lambda_m - \lambda)^{k_m'} det\,(\mathbf{C}_0 - \lambda\mathbf{I}_{n-r}),$$

where $\mathbf{I}_{n-r}$ denotes the $(n - r) \times (n - r)$ identity matrix. The roots of this polynomial must coincide with the characteristic roots of $\mathbf{A}$, since $\mathbf{A}$ and $\mathbf{C}$ are similar.

Now suppose that $k_1' < k_1$. This means that λ_1 is a root of the polynomial $det\,(\mathbf{C}_0 - \lambda\mathbf{I}_{n-r})$, i.e., λ_1 must be a characteristic root of the matrix $\mathbf{C}_0$. Therefore, there exists a nonnull vector $\mathbf{x}_0$ in $\mathbf{R}^{n-r}$ such that $\mathbf{C}_0\mathbf{x}_0 = \lambda_1\mathbf{x}_0$. Let $\mathbf{x}$ be the member of $\mathbf{R}^n$ whose first r components are zero and whose last $n - r$ components are the components of the vector $\mathbf{x}_0$. By construction, we now have

$$\mathbf{C}\mathbf{x} = \lambda_1\mathbf{x},$$

or, in other words,

$$\mathbf{A}\hat{\mathbf{B}}\mathbf{x} = \lambda_1\hat{\mathbf{B}}\mathbf{x}.$$

Hence, $\hat{\mathbf{B}}\mathbf{x}$ is a characteristic vector of $\mathbf{A}$, associated with the characteristic root λ_1. Therefore, by definition, $\hat{\mathbf{B}}\mathbf{x}$ is in the set S_1. However, by construction, $\hat{\mathbf{B}}\mathbf{x}$ is a linear combination of the last $n - r$ columns of $\hat{\mathbf{B}}$, which form a basis for $S^{\perp}$. Hence, $\hat{\mathbf{B}}\mathbf{x}$ must be a member of $S^{\perp}$ and, in particular, it must be orthogonal to every member of S_1. This is a contradiction: a nonnull vector is required to be a member of S_1 and, simultaneously, to be orthogonal to every member of S_1. Hence, we must have $k_1' \geq k_1$. By the same token, $k_j' \geq k_j$ for all members j of the set $\{1, 2, \ldots, m\}$. But since

$$\sum_{j=1}^{m} k_j' = r \leq n = \sum_{j=1}^{m} k_j,$$

we must have $k_j' = k_j$ for all j. ∎

Corollary 1 (Diagonalization of a Real Symmetric Matrix)

Every real symmetric matrix is orthogonally similar to a diagonal matrix.

PROOF:
Let $\mathbf{A}$ be a real symmetric $n \times n$ matrix, and construct the orthogonal matrix $\mathbf{B}$ as in the foregoing proof. Then, $\mathbf{B}'\mathbf{A}\mathbf{B}$ is a diagonal matrix. ∎

Corollary 2

Let $\mathbf{A}$ be a real symmetric $n \times n$ matrix, and let $\lambda_1, \lambda_2, \ldots, \lambda_s$ be the nonzero characteristic roots of $\mathbf{A}$, with corresponding multiplicities $k_1, k_2, \ldots, k_s$. Then,

$$rank\ \mathbf{A} = \sum_{i=1}^{s} k_i.$$

In other words, the rank of a real symmetric matrix is equal to the number of nonzero characteristic roots (counting multiplicities) of this matrix.

PROOF:
Left to the reader. ∎

PROBLEMS

15.1. Show that the product of the characteristic roots of a real symmetric matrix is equal to its determinant.

15.2. Let $\mathbf{A}$ be a real $n \times n$ matrix, with entries a_{ij}. The sum of the entries on the main diagonal of $\mathbf{A}$ is called the **trace** of $\mathbf{A}$, and it is denoted $tr\mathbf{A}$:

$$tr\mathbf{A} = a_{11} + a_{22} + \cdots + a_{nn}.$$

Show that the sum of the characteristic roots of a real symmetric matrix is equal to its trace. (Hint: One way to show this is to begin by showing that $tr\mathbf{AB} = tr\mathbf{BA}$.)

15.3. A real $n \times n$ matrix $\mathbf{A}$ is said to be **idempotent** iff $\mathbf{AA} = \mathbf{A}$. Show that if λ is a characteristic root of an idempotent matrix, then λ is either 0 or 1.

15.4. Let the following $n \times n$ matrix

$$\begin{bmatrix} 1 & \alpha & 0 & 0 & \ldots & 0 & 0 \\ \alpha & 1 & \alpha & 0 & \ldots & 0 & 0 \\ 0 & \alpha & 1 & \alpha & \ldots & 0 & 0 \\ 0 & 0 & \alpha & 1 & \ldots & 0 & 0 \\ \cdot & \cdot & \cdot & & \cdot & \cdot \\ \cdot & \cdot & \cdot & & \cdot & \cdot \\ \cdot & \cdot & \cdot & & \cdot & \cdot \\ 0 & 0 & 0 & 0 & & \alpha & 1 \end{bmatrix}$$

be denoted $\mathbf{A}_n$. Find the characteristic roots of $\mathbf{A}_n$ for $n = 1, 2, 3, 4, 5$.

15.5. Once again, consider the matrix $\mathbf{A}_n$ of Problem 15.4, and assume that α satisfies the inequality $-\frac{1}{2} \leq \alpha \leq \frac{1}{2}$. Let $det\ \mathbf{A}_n$ be denoted D_n, and define the vector $\mathbf{x}_n$ as follows:

$$\mathbf{x}_n = \begin{bmatrix} D_n \\ D_{n-1} \end{bmatrix}$$

for $n = 1, 2, \ldots$. $\mathbf{x}_n$ is, of course, a member of the space $\mathbf{R}^2$. Referring to Problem 14.6, page 142, find a real matrix $\mathbf{B}$ such that

$$\mathbf{x}_n = \mathbf{B}\mathbf{x}_{n-1}.$$

Find the characteristic roots of $\mathbf{B}$, and find the normalized characteristic vectors of $\mathbf{B}$. From your findings, can you deduce that D_n approaches zero as n becomes very large?

15.6. Let $\mathbf{x}_1, \mathbf{x}_2, \ldots, \mathbf{x}_m$ be members of the space $\mathbf{R}^n$. Define the **overall variance** of the vectors $\mathbf{x}_1, \mathbf{x}_2, \ldots, \mathbf{x}_m$ to be the quantity

$$(\mathbf{x}_1, \mathbf{x}_1) + (\mathbf{x}_2, \mathbf{x}_2) + \cdots + (\mathbf{x}_m, \mathbf{x}_m).$$

Show that there exist m vectors $\mathbf{u}_1, \mathbf{u}_2, \ldots, \mathbf{u}_m$ in $\mathbf{R}^n$ with the following three properties:

   (i) $\mathbf{u}_j$ is a linear combination of $\mathbf{x}_1, \mathbf{x}_2, \ldots, \mathbf{x}_m$ for $j = 1, 2, \ldots m$;
   (ii) The vectors $\mathbf{u}_1, \mathbf{u}_2, \ldots, \mathbf{u}_m$ are pairwise orthogonal;

(iii) The overall variance of $\mathbf{u}_1, \mathbf{u}_2, \ldots, \mathbf{u}_m$ is equal to the overall variance of $\mathbf{x}_1, \mathbf{x}_2, \ldots, \mathbf{x}_m$.

The vectors $\mathbf{u}_1, \mathbf{u}_2, \ldots, \mathbf{u}_m$ are sometimes referred to as the **principal components** associated with the vectors $\mathbf{x}_1, \mathbf{x}_2, \ldots, \mathbf{x}_m$. (Hint: Let $\mathbf{X}$ be the matrix whose columns are the vectors $\mathbf{x}_1, \mathbf{x}_2, \ldots, \mathbf{x}_m$. Diagonalize the matrix $\mathbf{X'X}$ and then make use of Problem 15.2.)

16   Quadratic Forms

Let Q be a real valued function (that is, a function with values in **R**) on $\mathbf{R}^n$. We say that Q is a **quadratic form** on $\mathbf{R}^n$ iff there exists a real $n \times n$ matrix **A** such that for every element **x** of $\mathbf{R}^n$,

$$Q(\mathbf{x}) = (\mathbf{x}, \mathbf{Ax}),$$

or, in the alternative notation for the inner product, such that

$$Q(\mathbf{x}) = \mathbf{x}'\mathbf{Ax}.$$

If Q is a quadratic form on $\mathbf{R}^n$, then, without loss of generality, we may assume that the matrix **A**, which enters the definition of Q, is a symmetric matrix. For suppose that Q is the quadratic form such that $Q(\mathbf{x}) = \mathbf{x}'\mathbf{Ax}$ for every $\mathbf{x} \in \mathbf{R}^n$. Let the matrix $\frac{1}{2}(\mathbf{A} + \mathbf{A}')$ be denoted **B**, and note that **B** is a symmetric matrix. Let $\hat{Q}$ be the quadratic form such that $\hat{Q}(\mathbf{x}) = \mathbf{x}'\mathbf{Bx}$ for every $\mathbf{x} \in \mathbf{R}^n$. Then,

$$\hat{Q}(\mathbf{x}) = \tfrac{1}{2}\mathbf{x}'(\mathbf{A} + \mathbf{A}')\mathbf{x}$$
$$= \tfrac{1}{2}[\mathbf{x}'\mathbf{Ax} + \mathbf{x}'\mathbf{A}'\mathbf{x}].$$

Now, $\mathbf{x}'\mathbf{A}\mathbf{x}$ can be looked upon as a real 1×1 matrix, so that its transpose is equal to itself. Hence,

$$\mathbf{x}'\mathbf{A}\mathbf{x} = [\mathbf{x}'\mathbf{A}\mathbf{x}]' = \mathbf{x}'\mathbf{A}'\mathbf{x},$$

so that

$$\begin{aligned}\hat{Q}(\mathbf{x}) &= \tfrac{1}{2}[\mathbf{x}'\mathbf{A}\mathbf{x} + \mathbf{x}'\mathbf{A}\mathbf{x}] \\ &= \mathbf{x}'\mathbf{A}\mathbf{x} \\ &= Q(\mathbf{x}).\end{aligned}$$

Hence, the quadratic forms Q and $\hat{Q}$, the latter involving a symmetric matrix, coincide.

Let Q be a quadratic form on $\mathbf{R}^n$. We say that Q is **positive definite** (respectively, **negative definite**) iff the inequality $Q(\mathbf{x}) > 0$ (respectively, $Q(\mathbf{x}) < 0$) holds for every nonnull member $\mathbf{x}$ of $\mathbf{R}^n$. We say that Q is **positive semidefinite** (respectively, **negative semidefinite**) iff the inequality $Q(\mathbf{x}) \geqq 0$ (respectively, $Q(\mathbf{x}) \leqq 0$) holds for every element $\mathbf{x}$ of $\mathbf{R}^n$. Our aim in this chapter is to formulate conditions on a real symmetric matrix $\mathbf{A}$ that ensure that the quadratic form Q, defined by $Q(\mathbf{x}) = \mathbf{x}'\mathbf{A}\mathbf{x}$, will be, say, positive definite (or that it will have any of the other properties just introduced). Conditions of this sort have applications in many areas, including, for example, the theory of maxima and minima.

Proposition 16.1

Let **A** *be a real symmetric* $n \times n$ *matrix, and let* Q *be the quadratic form such that* $Q(\mathbf{x}) = \mathbf{x}'\mathbf{A}\mathbf{x}$ *for every element* **x** *of* $\mathbf{R}^n$. *In addition, let* $\lambda_1, \lambda_2, \ldots, \lambda_n$ *be the characteristic roots of* **A**. *Then, the following four assertions are valid:*

(i) Q *is positive definite iff the characteristic roots* $\lambda_1, \lambda_2, \ldots, \lambda_n$ *are all positive;*

(ii) Q *is negative definite iff* $\lambda_1, \lambda_2, \ldots, \lambda_n$ *are all negative;*

(iii) Q *is positive semidefinite iff* $\lambda_1, \lambda_2, \ldots, \lambda_n$ *are all nonnegative;*

(iv) Q *is negative semidefinite iff* $\lambda_1, \lambda_2, \ldots, \lambda_n$ *are all nonpositive.*

PROOF:

By the proposition on the diagonalization of a real symmetric matrix, there exists an orthogonal matrix **C** such that

$$\mathbf{C}'\mathbf{A}\mathbf{C} = \mathbf{D}$$

where **D** is the diagonal matrix whose diagonal entries are the characteristic roots $\lambda_1, \lambda_2, \ldots, \lambda_n$. Since **C** is nonsingular, the statement

$$\mathbf{x}'\mathbf{A}\mathbf{x} > 0 \quad \text{for all} \quad \mathbf{x} \neq \mathbf{0}$$

is obviously equivalent to the statement

$$\mathbf{y}'\mathbf{C}'\mathbf{A}\mathbf{C}\mathbf{y} > 0 \quad \text{for all} \quad \mathbf{y} \neq \mathbf{0}.$$

But $\mathbf{C}'\mathbf{A}\mathbf{C} = \mathbf{D}$ and it is easy to verify that if **y** is a member of $\mathbf{R}^n$ with components $y_1, y_2, \ldots, y_n$, then

$$\mathbf{y}'\mathbf{D}\mathbf{y} = \sum_{i=1}^{n} \lambda_i y_i^2,$$

and this latter quantity is positive for all nonnull elements **y** of $\mathbf{R}^n$ iff the real numbers $\lambda_1, \lambda_2, \ldots, \lambda_n$ are all positive. This proves assertion (i). The other assertions are proved in a similar manner. ∎

Let **A** be an arbitrary real $n \times n$ matrix, and let k be any integer in the set $\{1, 2, \ldots, n\}$. Throughout the remainder of this chapter, the symbol $\mathbf{A}_k$ will be used to denote the $k \times k$ matrix that is obtained by deleting all but the first k columns and all but the first k rows of $\mathbf{A}$. Thus, if the typical entry of $\mathbf{A}$ is a_{ij}, then

$$\mathbf{A}_1 = a_{11}$$

$$\mathbf{A}_2 = \begin{bmatrix} a_{11} & a_{12} \\ a_{21} & a_{22} \end{bmatrix}$$

and so forth. The determinant of the matrix $\mathbf{A}_k$ is referred to as the k-th **principal minor** of $\mathbf{A}$.†

Before proceeding to the main proposition of this chapter, we need one more definition and a preliminary proposition. A real $n \times n$ matrix $\mathbf{T}$, with entries t_{ij}, is said to be **triangular** iff $t_{ij} = 0$ whenever $i > j$.

Proposition 16.2

Let $\mathbf{A}$, $\mathbf{B}$, and $\mathbf{C}$ be real $n \times n$ matrices. If $\mathbf{B}$ and $\mathbf{C}$ are triangular, then

$$(\mathbf{B}'\mathbf{A}\mathbf{C})_k = \mathbf{B}'_k \mathbf{A}_k \mathbf{C}_k$$

for $k = 1, 2, \ldots, n$.

PROOF:
Left to the reader. ∎

Proposition 16.3

Let $\mathbf{A}$ be a real symmetric $n \times n$ matrix, and for $k = 1, 2, \ldots, n$, let Δ_k be the k-th principal minor of $\mathbf{A}$. (In other words, define $\Delta_1 = \det \mathbf{A}_1$, $\Delta_2 = \det \mathbf{A}_2$, and so forth.) If $\Delta_k \neq 0$ for $k = 1, 2, \ldots, n$, then there exists a real $n \times n$ matrix $\mathbf{T}$ such that

$$\mathbf{T}'\mathbf{A}\mathbf{T} = \begin{bmatrix} \Delta_1 & 0 & 0 & \ldots & 0 \\ 0 & \dfrac{\Delta_2}{\Delta_1} & 0 & \ldots & 0 \\ 0 & 0 & \dfrac{\Delta_3}{\Delta_2} & \ldots & 0 \\ \cdot & \cdot & \cdot & & \cdot \\ \cdot & \cdot & \cdot & & \cdot \\ \cdot & \cdot & \cdot & & \cdot \\ 0 & 0 & 0 & \ldots & \dfrac{\Delta_n}{\Delta_{n-1}} \end{bmatrix}$$

† The determinant $\det \mathbf{A}_k$ should really be called the k-th *upper corner* principal minor. However, as a matter of convenience, we shall omit the phrase "upper corner."

PROOF:

By hypothesis, $\Delta_1 \neq 0$. In other words, $a_{11} \neq 0$. We can therefore perform a series of elementary operations of the third type (see Chapter 11) on the rows of $\mathbf{A}$ with the result that all the entries in the first column of $\mathbf{A}$, except a_{11}, will reduce to 0. As we already know, performing elementary operations on the rows of $\mathbf{A}$ is equivalent to premultiplying $\mathbf{A}$ by a series of elementary matrices. Specifically, it is easy to see that reducing the entries in the first column of $\mathbf{A}$ to 0 is equivalent to premultiplying $\mathbf{A}$ by the transpose of the matrix $\mathbf{T}^{(1)}$, whose entries, $t_{ij}^{(1)}$, are given by

$$
\begin{aligned}
t_{ij}^{(1)} &= \delta_{ij} & &\text{if } i \neq 1 \\
&= 1 & &\text{if } i = 1 \text{ and } j = 1 \\
&= -a_{1j}/a_{11} & &\text{if } i = 1 \text{ and } j > 1,
\end{aligned}
$$

where δ_{ij} is the Kronecker delta. Now, since $\mathbf{A}$ is a symmetric matrix, postmultiplying $\mathbf{A}$ by $\mathbf{T}^{(1)}$ does to the columns of $\mathbf{A}$ exactly what premultiplying by $\mathbf{T}^{(1)'}$ does to the rows of $\mathbf{A}$. In other words, the matrix

$$
\mathbf{T}^{(1)'}\mathbf{A}\mathbf{T}^{(1)}
$$

is the matrix obtained after all the entries in the first column of $\mathbf{A}$ (except a_{11}) are reduced to 0 by elementary operations on the rows of $\mathbf{A}$, and after all the entries in the first row of $\mathbf{A}$ (except a_{11}) are reduced to 0 by elementary operations on the columns of $\mathbf{A}$.

Let the matrix $\mathbf{T}^{(1)'}\mathbf{A}\mathbf{T}^{(1)}$ be denoted $\mathbf{A}^{(2)}$, and let its typical entry be denoted $a_{ij}^{(2)}$. If $a_{22}^{(2)} \neq 0$, then we can repeat the foregoing process and reduce the entries in the second column and in the second row of $\mathbf{A}^{(2)}$ (except for the entry $a_{22}^{(2)}$ itself) to 0 by means of elementary operations of the third type. In other words, assuming for the moment that $a_{22}^{(2)} \neq 0$, let us define a real $n \times n$ matrix $\mathbf{T}^{(2)}$, with entries $t_{ij}^{(2)}$, as follows:

$$
\begin{aligned}
t_{ij}^{(2)} &= \delta_{ij} & &\text{if } i \neq 2 \\
&= 0 & &\text{if } i = 2 \text{ and } j = 1 \\
&= 1 & &\text{if } i = 2 \text{ and } j = 2 \\
&= -a_j^{(2)}/a_{22}^{(2)} & &\text{if } i = 2 \text{ and } j > 2.
\end{aligned}
$$

Then, the matrix $\mathbf{T}^{(2)'}\mathbf{A}^{(2)}\mathbf{T}^{(2)}$ is a matrix with zeros in its first two columns and in its first two rows, except on the main diagonal. Let this matrix be denoted $\mathbf{A}^{(3)}$, and let its typical entry be denoted $a_{ij}^{(3)}$. If $a_{33}^{(3)} \neq 0$, we can proceed to reduce the entries in the third column and in the third row of $\mathbf{A}^{(3)}$ to 0, by elementary operations of the third type.

To make a long story somewhat shorter, what we are doing here is defining a sequence of matrices $\mathbf{A}^{(1)}, \mathbf{A}^{(2)}, \ldots, \mathbf{A}^{(n)}$ by the following recursive relationship:

$$\mathbf{A}^{(1)} = \mathbf{A}$$

$$\mathbf{A}^{(m+1)} = \mathbf{T}^{(m)'}\mathbf{A}^{(m)}\mathbf{T}^{(m)} \quad \text{for} \quad m = 1, 2, \ldots, n-1$$

where $\mathbf{T}^{(m)}$ is a real $n \times n$ matrix defined as follows: If $a_{ij}^{(m)}$ is the typical entry of $\mathbf{A}^{(m)}$, then $t_{ij}^{(m)}$, the typical entry of $\mathbf{T}^{(m)}$, is given by

$$
\begin{aligned}
t_{ij}^{(m)} &= \delta_{ij} & &\text{if} \quad i \neq m \\
&= 0 & &\text{if} \quad i = m \quad \text{and} \quad j < m \\
&= 1 & &\text{if} \quad i = m \quad \text{and} \quad j = m \\
&= -a_{mj}^{(m)}/a_{mm}^{(m)} & &\text{if} \quad i = m \quad \text{and} \quad j > m.
\end{aligned}
$$

Assume that we have managed to carry out $m-1$ steps of this procedure, so that we have reached the matrix $\mathbf{A}^{(m)}$, where m is some integer in the set $\{1, 2, \ldots, n\}$. Obviously, we can take the next step, i.e., proceed from the matrix $\mathbf{A}^{(m)}$ to the matrix $\mathbf{A}^{(m+1)}$, iff the entry $a_{mm}^{(m)}$ is not zero. We shall now show that, given that we have been able to reach the matrix $\mathbf{A}^{(m)}$, we can proceed to $\mathbf{A}^{(m+1)}$ iff $\Delta_m \neq 0$. In other words, we intend to show that

$$a_{mm}^{(m)} \neq 0 \quad \text{iff} \quad \Delta_m \neq 0.$$

In the process, we shall also show that $a_{mm}^{(m)} = \Delta_m/\Delta_{m-1}$, which will bring us to the desired result.

Recall that going from $\mathbf{A}^{(m-1)}$ to $\mathbf{A}^{(m)}$ means reducing all the entries in column $m-1$ of $\mathbf{A}^{(m-1)}$ and all the entries in row $m-1$ of $\mathbf{A}^{(m-1)}$ to 0 by means of elementary operations of the third type. Thus, $\mathbf{A}^{(m)}$ is of the form

$$
\mathbf{A}^{(m)} =
\begin{bmatrix}
a_{11}^{(1)} & 0 & 0 & \cdots & 0 & 0 & \cdots & 0 \\
0 & a_{22}^{(2)} & 0 & \cdots & 0 & 0 & \cdots & 0 \\
0 & 0 & a_{33}^{(3)} & \cdots & 0 & 0 & \cdots & 0 \\
\cdot & \cdot & \cdot & & \cdot & \cdot & & \cdot \\
\cdot & \cdot & \cdot & & \cdot & \cdot & & \cdot \\
\cdot & \cdot & \cdot & & \cdot & \cdot & & \cdot \\
0 & 0 & 0 & \cdots & a_{mm}^{(m)} & a_{m,m+1}^{(m)} & \cdots & a_{mn}^{(m)} \\
0 & 0 & 0 & \cdots & a_{m+1,m}^{(m)} & a_{m+1,m+1}^{(m)} & \cdots & a_{m+1,n}^{(m)} \\
\cdot & \cdot & \cdot & & \cdot & \cdot & & \cdot \\
\cdot & \cdot & \cdot & & \cdot & \cdot & & \cdot \\
\cdot & \cdot & \cdot & & \cdot & \cdot & & \cdot \\
0 & 0 & 0 & \cdots & a_{nm}^{(m)} & a_{n,m+1}^{(m)} & \cdots & a_{nn}^{(m)}
\end{bmatrix}
$$

Next, we claim that the principal minors of $\mathbf{A}^{(m)}$ coincide with the corresponding principal minor of $\mathbf{A}^{(m-1)}$. To see this, observe that the matrices $\mathbf{T}^{(1)}, \mathbf{T}^{(2)}, \mathbf{T}^{(3)}, \ldots$ are all triangular. Therefore, by Proposition 16.2,

$$\mathbf{A}_k^{(m)} = \mathbf{T}_k^{(m-1)\prime} \mathbf{A}_k^{(m-1)} \mathbf{T}_k^{(m-1)} \quad \text{for} \quad k = 1, 2, \ldots, n.$$

Hence,

$$det\ \mathbf{A}_k^{(m)} = (det\ \mathbf{T}_k^{(m-1)})^2\ det\ \mathbf{A}_k^{(m-1)}.$$

But the matrices $\mathbf{T}^{(1)}, \mathbf{T}^{(2)}, \ldots$ are not just triangular. Their diagonal entries are all equal to 1, and it is immediately clear that the determinant of a triangular matrix whose diagonal entries are all equal to 1 is itself equal to 1. In other words,

$$det\ \mathbf{T}_k^{(m-1)} = 1$$

for $m = 2, 3, \ldots, n$ and for $k = 1, 2, \ldots, n$. So, we have

$$det\ \mathbf{A}_k^{(m)} = det\ \mathbf{A}_k^{(m-1)} \quad \text{for} \quad k = 1, 2, \ldots, n.$$

By the same token, $det\ \mathbf{A}_k^{(m-1)} = det\ \mathbf{A}_k^{(m-2)}$, and so on. In short, the transformation which carries the matrix $\mathbf{A}^{(m-1)}$ into the matrix $\mathbf{A}^{(m)}$ is *minor-preserving*, and we have, for any k and m in the set $\{1, 2, \ldots, n\}$,

$$det\ \mathbf{A}_k^{(m)} = \Delta_k.$$

Now, for $k = 1, 2, \ldots, m$, the minors $det\ \mathbf{A}_k^{(m)}$ are very simple to calculate, because of the special structure of the matrix $\mathbf{A}^{(m)}$. Specifically, we have

$$det\ \mathbf{A}_1^{(m)} = a_{11}^{(1)}$$
$$det\ \mathbf{A}_2^{(m)} = a_{11}^{(1)} a_{22}^{(2)}$$
$$\vdots$$
$$det\ \mathbf{A}_m^{(m)} = a_{11}^{(1)} a_{22}^{(2)} \cdots a_{mm}^{(m)}.$$

Combining this with our previous finding, we get

$$a_{11}^{(1)} = \Delta_1$$
$$a_{22}^{(2)} = \Delta_2/\Delta_1$$
$$\vdots$$
$$a_{mm}^{(m)} = \Delta_m/\Delta_{m-1}.$$

Hence, given that we have reached the matrix $\mathbf{A}^{(m)}$, we can proceed to the matrix $\mathbf{A}^{(m+1)}$ iff $\Delta_m \neq 0$. But by hypothesis, $\Delta_m \neq 0$ for $m = 1$, $2, \ldots, n$, and therefore the process which we have described here can proceed smoothly all the way, until the matrix $\mathbf{A}^{(n)}$ is reached. Furthermore, in the course of showing this we have, in fact, established that the final matrix, $\mathbf{A}^{(n)}$, will be of the form

$$
\begin{bmatrix}
\Delta_1 & 0 & 0 & \cdots & 0 \\
0 & \dfrac{\Delta_2}{\Delta_1} & 0 & \cdots & 0 \\
0 & 0 & \dfrac{\Delta_3}{\Delta_2} & \cdots & 0 \\
\cdot & \cdot & \cdot & & \cdot \\
\cdot & \cdot & \cdot & & \cdot \\
\cdot & \cdot & \cdot & & \cdot \\
0 & 0 & 0 & \cdots & \dfrac{\Delta_n}{\Delta_{n-1}}
\end{bmatrix}
$$

so that the matrix $\mathbf{T}$ that we have been seeking is given by the product $\mathbf{T}^{(1)}\mathbf{T}^{(2)} \ldots \mathbf{T}^{(n-1)}$. ∎

Proposition 16.4

Let $\mathbf{A}$ be a real symmetric $n \times n$ matrix, and let its principal minors be denoted $\Delta_1, \Delta_2, \ldots, \Delta_n$. Consider the quadratic form Q such that $Q(\mathbf{x}) = \mathbf{x}'\mathbf{A}\mathbf{x}$ for every member $\mathbf{x}$ of $\mathbf{R}^n$. Q is positive definite iff $\Delta_k > 0$ for $k = 1, 2, \ldots, n$.

PROOF:

We must establish two implications. First, if $Q(\mathbf{x}) > 0$ for all nonnull vectors $\mathbf{x}$ in $\mathbf{R}^n$, then $\Delta_k > 0$ for $k = 1, 2, \ldots, n$. Second, if $\Delta_k > 0$ for $k = 1, 2, \ldots, n$, then $Q(\mathbf{x}) > 0$ for all nonnull $\mathbf{x}$ in $\mathbf{R}^n$. In *both* of these we are entitled to assume that $\Delta_k \neq 0$ for $k = 1, 2, \ldots, n$. In the case of the second implication, this is trivial. As for the first implication, suppose that $\Delta_k = 0$ for some k in the set $\{1, 2, \ldots, n\}$. This means that the submatrix $\mathbf{A}_k$ is singular, so that there exists a nonnull vector, say $\mathbf{w}$, in $\mathbf{R}^k$ such that $\mathbf{A}_k\mathbf{w} = \mathbf{0}$. This certainly implies that $\mathbf{w}'\mathbf{A}_k\mathbf{w} = 0$. Letting $\mathbf{x}$ be the vector in $\mathbf{R}^n$ whose first k components are the components of $\mathbf{w}$ and whose last $n - k$ components are 0, it is easy to see that $\mathbf{x}'\mathbf{A}\mathbf{x} = \mathbf{w}'\mathbf{A}_k\mathbf{w} = 0$. But, by construction, $\mathbf{x}$ is nonnull, and therefore Q cannot be positive definite.

Given the knowledge that $\Delta_k \neq 0$ for $k = 1, 2, \ldots, n$, we are now in a position to make use of Proposition 16.3. Let $\mathbf{T}$ be the matrix such that

$$\mathbf{T'AT} = \begin{bmatrix} \Delta_1 & 0 & \cdots & 0 \\ 0 & \dfrac{\Delta_2}{\Delta_1} & \cdots & 0 \\ \vdots & \vdots & & \vdots \\ \cdot & \cdot & & \cdot \\ \cdot & \cdot & & \cdot \\ 0 & 0 & \cdots & \dfrac{\Delta_n}{\Delta_{n-1}} \end{bmatrix}$$

and define a new quadratic form $\hat{Q}$ as follows:

$$\hat{Q}(\mathbf{x}) = \mathbf{x}'\mathbf{T}'\mathbf{AT}\mathbf{x}$$

for all members $\mathbf{x}$ of $\mathbf{R}^n$. Recall that $\mathbf{T}$ is the product of elementary matrices, which implies that it is nonsingular. Therefore, the quadratic form Q is positive definite iff the quadratic form $\hat{Q}$ is positive definite. But clearly, $\hat{Q}$ is positive definite iff all the diagonal entries of $\mathbf{T}'\mathbf{AT}$ are positive, i.e., iff $\Delta_k > 0$ for $k = 1, 2, \ldots, n$. ∎

Proposition 16.5

Let $\mathbf{A}$ be a real symmetric $n \times n$ matrix, and let its principal minors be denoted $\Delta_1, \Delta_2, \ldots, \Delta_n$. Consider the quadratic form Q such that $Q(\mathbf{x}) = \mathbf{x}'\mathbf{A}\mathbf{x}$ for every member $\mathbf{x}$ of $\mathbf{R}^n$. Q is negative definite iff $(-1)^k \Delta_k > 0$ for $k = 1, 2, \ldots, n$.

PROOF:

Q is negative definite iff $-Q$ is positive definite, i.e., iff $\mathbf{x}'(-\mathbf{A})\mathbf{x} > 0$ for every nonnull member $\mathbf{x}$ of $\mathbf{R}^n$. By Proposition 16.4, the latter is true iff the principal minors of $-\mathbf{A}$ are all positive. But, by the properties of determinants, we have, for any real $k \times k$ matrix $\mathbf{B}$,

$$det \, (-\mathbf{B}) = (-1)^k \, det \, \mathbf{B}.$$

Hence, the k-th principal minor of $-\mathbf{A}$ is positive iff $(-1)^k \Delta_k > 0$. ∎

We conclude by noting that the conditions of Proposition 16.4 and Proposition 16.5 cannot be converted, by changing strict inequalities to weak inequalities, into necessary and sufficient conditions for positive semidefiniteness and negative semidefiniteness. More precisely, if $\mathbf{A}$ is a real symmetric $n \times n$ matrix whose upper-corner principal minors are denoted $\Delta_1, \Delta_2, \ldots, \Delta_n$, then the statement

$$\Delta_k \geqq 0 \quad \text{for} \quad k = 1, 2, \ldots, n$$

is a necessary condition, but not a sufficient condition, for the quadratic form Q such that $Q(\mathbf{x}) = \mathbf{x}'\mathbf{A}\mathbf{x}$ to be positive semidefinite. Similarly, $(-1)^k \Delta_k \geqq 0$ for $k = 1, 2, \ldots, n$ is a necessary condition, but not a sufficient condition for Q to be negative semidefinite. In fact, a necessary and sufficient condition for Q to be positive semidefinite, is that *all* the principal minors of $\mathbf{A}$, not only the upper-corner principal minors, be nonnegative. (A similar condition may be stated for negative semidefiniteness.) A principal minor of $\mathbf{A}$ that is not necessarily an upper-corner principal minor is simply the determinant of any matrix obtained by first deleting from $\mathbf{A}$ certain rows, say the rows $i_1, i_2, \ldots, i_k$, and then deleting the columns $i_1, i_2, \ldots, i_k$ as well. Thus, it is important to remember that throughout the foregoing discussion we have been using the phrase "principal minor" in the sense of "upper-corner principal minor."

PROBLEMS

16.1. Prove Proposition 16.2.

16.2. Let $\mathbf{A}$ be a real $m \times n$ matrix with rank n. Show that the quadratic form Q, such that $Q(\mathbf{x}) = \mathbf{x}'\mathbf{A}'\mathbf{A}\mathbf{x}$ for every element $\mathbf{x}$ of $\mathbf{R}^n$, is positive definite.

16.3. Let A be a real symmetric $n \times n$ matrix with entries a_{ij}, and consider the quadratic form Q such that $Q(\mathbf{x}) = \mathbf{x}'\mathbf{A}\mathbf{x}$ for every element $\mathbf{x}$ of $\mathbf{R}^n$. Let M be the subspace of $\mathbf{R}^n$ which consists of all members $\mathbf{x}$ of $\mathbf{R}^n$ whose first component is 0. Show that $Q(\mathbf{x}) > 0$ for every nonnull member of M iff the principal minors of the matrix

$$\begin{bmatrix} 0 & 1 & 0 & 0 & \ldots & 0 \\ 1 & a_{11} & a_{12} & a_{13} & \ldots & a_{1n} \\ 0 & a_{21} & a_{22} & a_{23} & \ldots & a_{2n} \\ 0 & a_{31} & a_{32} & a_{33} & \ldots & a_{3n} \\ \cdot & \cdot & \cdot & \cdot & & \cdot \\ \cdot & \cdot & \cdot & \cdot & & \cdot \\ \cdot & \cdot & \cdot & \cdot & & \cdot \\ 0 & a_{n1} & a_{n2} & a_{n3} & \ldots & a_{nn} \end{bmatrix}$$

are all negative (except, of course, for the first principal minor, which is 0).

16.4. Let $\langle \mathbf{x}_1, \mathbf{x}_2, \ldots, \mathbf{x}_n \rangle$ be a basis for $\mathbf{R}^n$, and let Q be a quadratic form on $\mathbf{R}^n$. Is the following assertion true: Q is positive definite iff $Q(\mathbf{x}_i) > 0$ for $i = 1, 2, \ldots, n$. Give a proof or a counterexample.

16.5. Find an example of a real symmetric matrix $\mathbf{A}$ whose upper-corner principal minors are all nonnegative, such that the quadratic form associated with **A** is not positive semidefinite.

16.6. Let **A** be a real symmetric $n \times n$ matrix. Show that if the quadratic form associated with **A** is negative definite, then the entries on the main diagonal of **A** are negative.

Index

A

Addition, 24, 32
Adjoint, 140
Algebra of square matrices, 85
Alternating multilinear form, 127
Antisymmetric relation, 12
Associative binary operation, 20
Asymmetric relation, 12
Augmented matrix of a linear equation, 96

B

Basis, 49
Binary operation, 20
Binary relation, 10

Linear equation, 92
Linear inequality, 106
Linear operator, 37
Linear programming, 49
Linear space, 35
Linear subspace, 36
Linear transformation, 37, 68
Linearly independent set, 47

M

Mapping, 16
Matrix, 71
Matrix multiplication, 72
Matrix of a linear transformation, 73
Matrix of a permutation, 124
Member of a set, 3
Minimal generating subset, 49
Minor, 159
Multilinear form, 126
Multiplication, 24

N

n-ary relation, 10
n-linear form, 126
n-tuple, 10
n-tuple spaces, 56
Natural matrix of a transformation, 73
Negative definite quadratic form, 157
Nonnegative vector, 107
Nonsingular matrix, 87
Null element, 33
Null matrix, 88
Null set, 4
Null space, 38
Nullity, 70

R

S